STRUCTURE AND BONDING

Volume 32

With 31 Figures and 16 Tables

Springer-Verlag
Berlin Heidelberg GmbH 1977

ISBN 978-3-662-15480-9 ISBN 978-3-540-37546-3 (eBook)
DOI 10.1007/978-3-540-37546-3

Library of Congress Catalog Card Number 67-11280

Originally published by Springer-Verlag Berlin Heidelberg New York in 1977.
Softcover reprint of the hardcover 1st edition 1977

Typesetting: R. & J. Blank, München.

Contents

STRUCTURE AND BONDING is issued at irregular intervals, according to the material received. With the acceptance for publication of a manuscript, copyright of all countries is vested exclusively in the publisher. Only papers not previously published elsewhere should be submitted. Likewise, the author guarantees against subsequent publication elsewhere. The text should be as clear and concise as possible, the manuscript written on one side of the paper only. Illustrations should be limited to those actually necessary.

Manuscripts will be accepted by the editors:

SPRINGER-VERLAG

D-6900 Heidelberg 1
P. O. Box 105280
Telephone (06221) 487·1
Telex 04-61723

D-1000 Berlin 33
Heidelberger Platz 3
Telephone (030) 822001
Telex 01-83319

SPRINGER-VERLAG
NEW YORK INC.

175, Fifth Avenue
New York, N.Y. 10010
Telephone 673-2660

Recent Results in the Chemistry of Transition Metal Clusters with Organic Ligands

Heinrich Vahrenkamp

Chemisches Laboratorium der Universität Freiburg, Germany

Table of Contents

1. Introduction

The political justification for transition metal cluster chemistry is the assumption that clusters are models in which metallic properties may be more easily studied than in the metals themselves. These properties include electronic phenomena such as color and conductivities as well as surface phenomena, such as atom arrangements and catalytic activities. Thus, there are two main lines of cluster research. The more academic line leads to the search for new types of clusters and their structure and bonding, whereas the more practical line leads to the investigation of reactivities with the hope that clusters may open catalytic pathways that neither plain metals nor mononuclear catalysts can provide. The interdependence of both lines is obvious.

As in other fields of chemistry, the literature on metal cluster compounds has grown exponentially. During the last five years as many papers have been published as in the ten years before. While many of them extend areas of research that had been discovered before, others definitely have advanced the field. Here the synthesis of "big" clusters with more than six metal atoms, some theoretical approaches to the bonding in clusters, the discovery of the fluxional character of many clusters, and several new and generally applicable reactions of clusters, primarily with unsaturated organic compounds, deserve mentioning. These things were hardly known at the beginning of this decade (*231, 241*). A progress report therefore seems appropriate.

The number of researchers dealing with metal clusters is still growing, too. This results partly from the practical aspects mentioned above, but mostly from the fact that X-ray and mass spectrometry instrumentation, the two essential tools for cluster research, are becoming increasingly available. At the same time, improved syntheses for the basic clusters have been developed, saving time (though not money because of ever increasing noble metal prices) for their investigation. And finally, it has been substantiated that many clusters survive drastic conditions, allowing reactions that had not been taken into consideration before.

It can be stated that metal cluster chemistry has established itself as an important branch of the science. The purpose of this review is to underline this by elaborating on the wealth of results now available. It cannot be overlooked, however, that it is still too early for substantial generalizations or predictions in this field of chemistry.

2. General Topics

In this chapter a few things are to be mentioned which hold true for several groups of clusters, whereas the next chapter will deal with specific compounds and their properties. Due to the above-mentioned limits on generalizations this chapter, which ought to to be the main part of a critical review, cannot go into much depth at present.

2.1. Scope of Review

Definitions of a metal atom cluster compound have been given (*109, 233, 241, 316*). Accordingly, in this review a cluster will be considered as a compound containing at least three metal atoms connected by metal-metal interactions in a triangular or polyhedral array.

The latest review trying to cover the whole field of transition metal cluster compounds was written five years ago (*241*). The literature expansion since then has rendered it difficult to repeat this task. It was therefore decided to limit the topic to clusters containing organic ligands. This eliminates the triangular and octahedral metal halogen clusters of niobium, tantalum, molybdenum, tungsten, and rhenium, of which recent publications (*200, 271*) provide access to previous results. Clusters consisting of or containing main group or post-transition metals which may have structures similar to those adapted by transition metal clusters (*316*) will also not be considered.

In certain classes of polynuclear complexes it is difficult to decide whether they fall under the definition of a cluster or not. This is because the term "metal-metal interaction" is flexible. Somewhat arbitrary decisions had to be reached, therefore, in these cases: Not included in this review are the oligomeric copper halide (organic ligand) complexes which have yielded quite a variety of structural types (*73, 94*). Also the iron-sulfur oligomers of biological significance will not be considered, of which a brief review has appeared (*207*). Interestingly, a recent X_α–SCF–MO calculation (*401*) has revealed metal-metal interactions in these complexes despite the relatively long iron-iron distances. Finally, there is a series of cubane-like complexes of four metal and four bridging atoms in which metal-metal bonding seems to be absent (*18, 361*) and which therefore fall outside the scope of this review.

The already voluminous review literature on clusters will be considered as a basis for this review. The topics treated so far are clusters in general (*109, 241*) and in connection with metal-metal bonding (*30, 338, 380*); special types of clusters like those with π-acceptor ligands (*231*), hydrides (*233*), carbonyls (*85, 86*) or methinyl tricobalt enneacarbonyls (*313, 317*); properties of clusters like structures (*56, 316*), fluxionality (*110*), mass spectra (*226*), vibrational spectra (*365*), and redox behavior (*292*). Clusters have been treated in the context of metal carbonyls (*3, 4*), metal sulfur complexes (*2, 381*), and in relation to coordination polyhedra (*297*). Reviews on the chemistry of rhenium (*273, 332*) and the organometallic chemistry of ruthenium and osmium (*376*) or copper, silver, and gold (*306*) have made special reference to clusters. And finally, a valuable source of information are the Specialist Periodical Reports on Organometallic Chemistry by the Chemical Society with chapters on metal carbonyls (*5*), metal-metal bonded complexes (*105, 115*), and on substitution reactions of complexes (*153, 290*).

This review will cover the literature from 1971 to 1975 and deal with syntheses, properties, and reactions of transition metal clusters with organic ligands. Reactions that destroy clusters will not be considered. Completeness is attempted but not guaranteed.

Outside this chapter on general aspects of structure, bonding, syntheses, and reactivities the material will be presented according to the position of the metals in the periodic table. For heteronuclear clusters the compounds will be treated as belonging to the metal later in the periodic table. Compounds mentioned in this review without giving a reference have been dealt with in former reviews.

The common abbreviations are used: M = metal, L = ligand, R = organic group, Me = methyl, Et = ethyl, Pr = propyl, Bu = butyl, Ar = aromatic group, Ph = phenyl, Cp = cyclopentadienyl, Hal = halogen, X = anionic substituent. Short straight lines around metal atoms denote carbonyl ligands omitted for the sake of clarity.

2.2. Structural Aspects

Unusual Structures. Former reviews treating clusters according to their basic structures (*241, 316*) have revealed that the most common metal atom arrangements are triangle, tetrahedron, and octahedron. Relatively few other oligonuclear species were observed. This situation has changed during the last five years, and unusual structures with four to six metal atoms and a number of clusters with more than six metal atoms have been described.

Among the tetranuclear clusters there exists the full range of possible structures now between the tetrahedron and the rectangle. The "open" tetrahedron with the

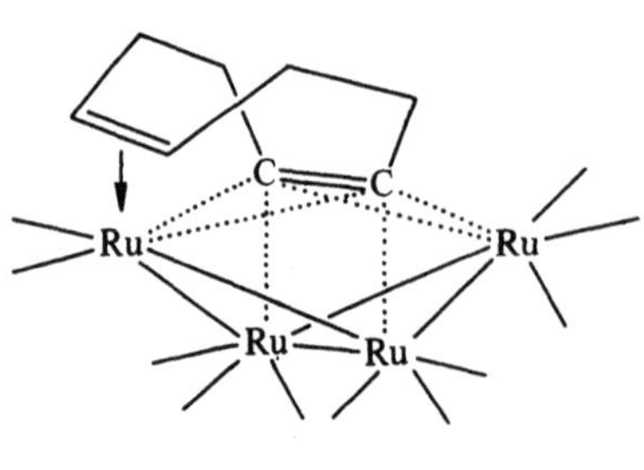

1

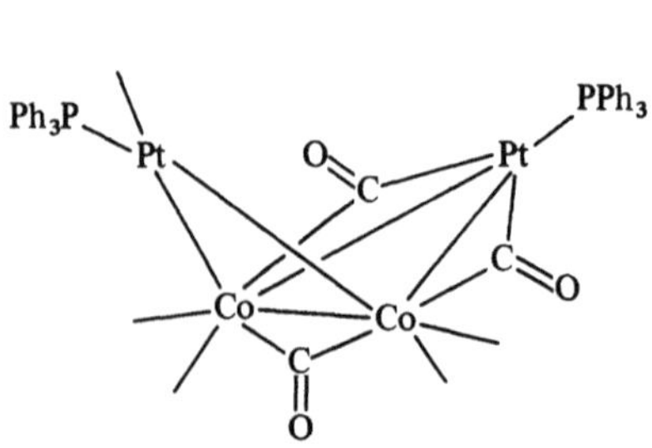

2

butterfly arrangement of metal atoms, which was formerly observed in cobalt acetylene complexes, has been found in two ruthenium olefin compounds $Ru_4(CO)_{10}$-$(C_{12}H_{16})$ (*36*) and $Ru_4(CO)_{11}(C_8H_{10})$ [*1*] (*283*) as well as in the heteronuclear carbonyl cluster $Pt_2Co_2(CO)_8(PPh_3)_2$ [*2*] (*174*).

A wider open butterfly angle is found in the tetrameric arylcopper [*3*] (*194, 263*), whereas a similar alkylcopper [*4*] (*225, 269*) has a square arrangement of copper atoms.

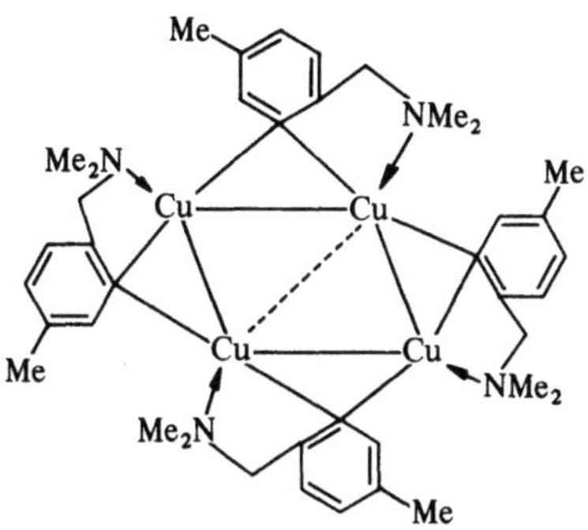

3

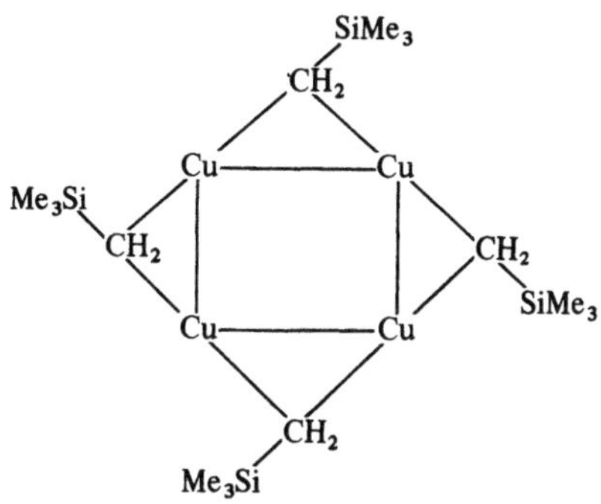

4

A rhombus of metal atoms as suggested for $HReOs_3(CO)_{15}$ [*5*] (*249*) in analogy to the known structure of $Re_4(CO)_{16}^{2-}$ may also be considered as resulting from a tetrahedron, while a rectangle as suggested for $HMOs_3(CO)_{16}$ (M = Mn, Re) (*249*) or found for $Co_4(CO)_{10}X_2$ [*6*] (*334*) (X = S, Se, PPh) is an independent and new type of cluster.

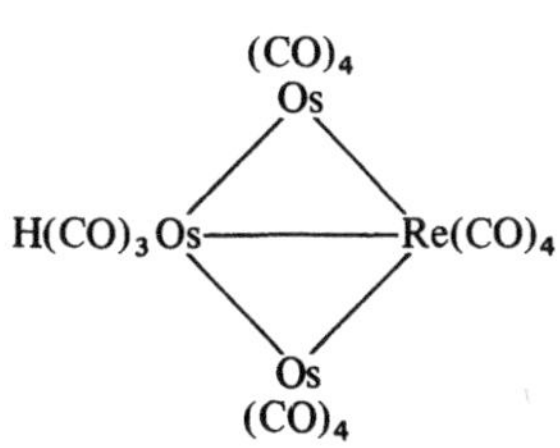

5

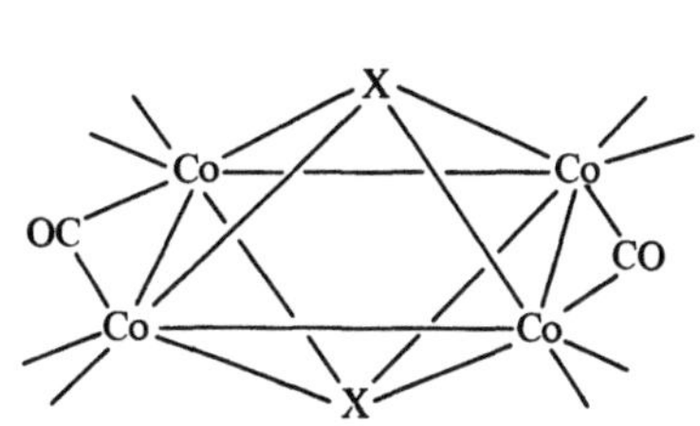

6

Yet another arrangement of four metal atoms, a triangle with one external metal, has been found for $H_4Re_4(CO)_{15}^{2-}$ [*7*] (*17*) and has been suggested as an alternative [*8*] for $HMOs_3(CO)_{16}$ (*249*).

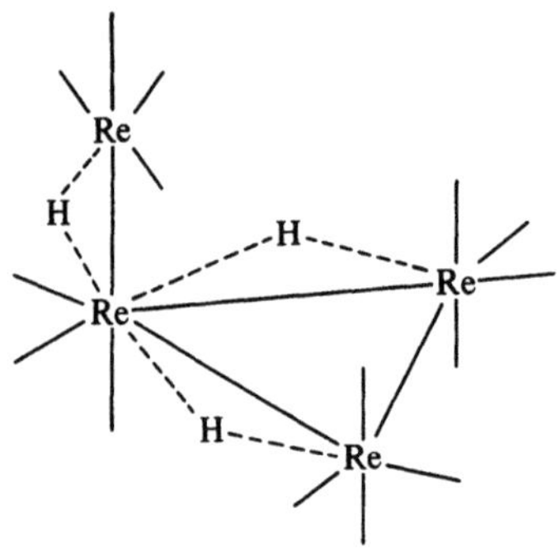

7

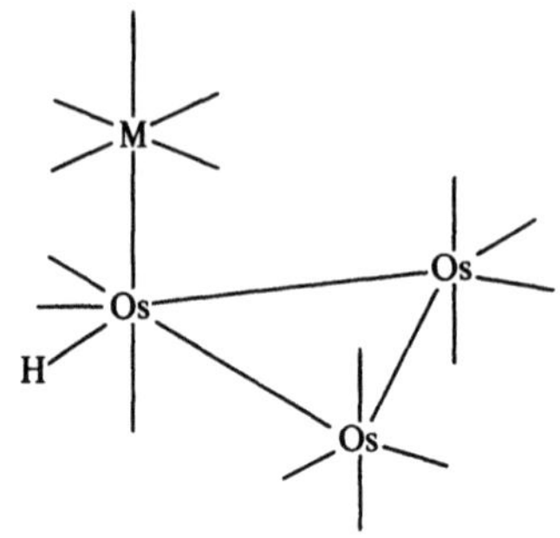

8

There are still only a small number of pentanuclear clusters, and only two structural types are known. These are the tetragonal pyramid as in $Fe_5C(CO)_{15}$, $Ru_5C(CO)_{15}$ (*152*), and $Os_5C(CO)_{15}$ (*151, 152*) or in the recently prepared $Fe_5C(CO)_{14}^{2-}$ (*214*), for which structure [*9*] was suggested. The metal atom arrangement of a trigonal bipyramid has been proved for two cases, namely $Ni_5(CO)_{12}^{2-}$ [*10*] (*274*) and $M_2Ni_3(CO)_{16}^{2-}$ (M = Cr, Mo, W) (*333*).

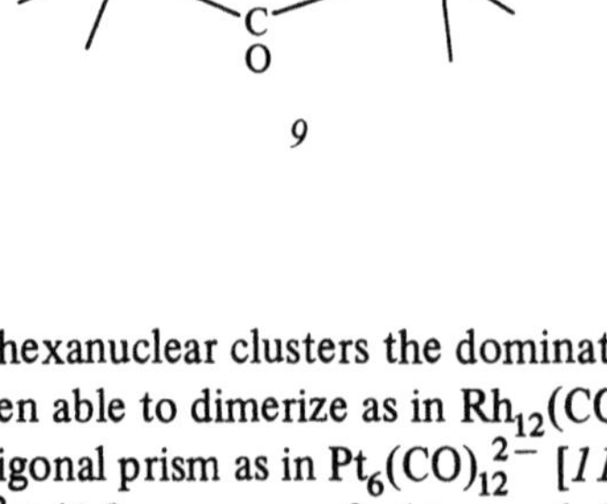

9

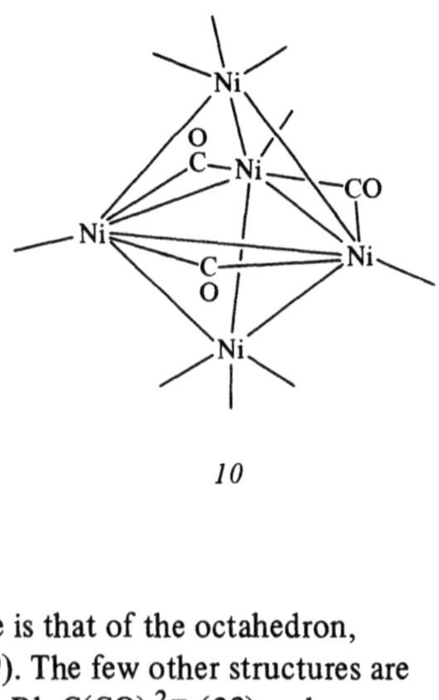

10

In the hexanuclear clusters the dominating shape is that of the octahedron, which is even able to dimerize as in $Rh_{12}(CO)_{30}^{2-}$ (*90*). The few other structures are that of a trigonal prism as in $Pt_6(CO)_{12}^{2-}$ [*11*] (*12*) or $Rh_6C(CO)_{15}^{2-}$ (*22*) and $Co_6C(CO)_{15}^{2-}$ (*14*), one case of a bicapped rhombus in $Mo_2Ni_4(CO)_{14}^{2-}$ (*333*), and one case of a bicapped tetrahedron in $Os_6(CO)_{18}$ [*12*] (*284*). The reasons for certain structures of hexanuclear clusters have been discussed in terms of several bonding approaches and will be treated in Chapter 2.3.

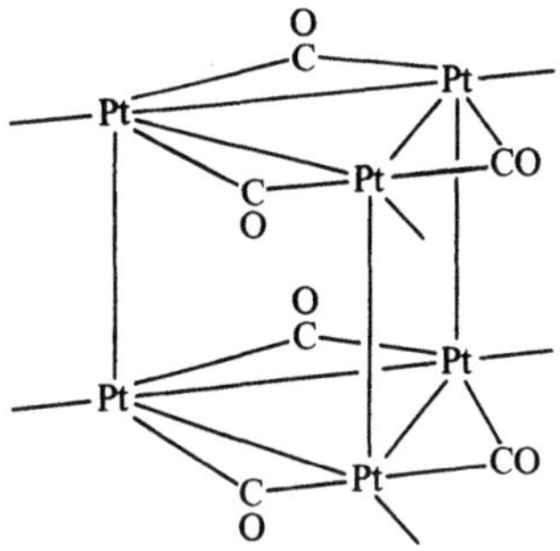

11

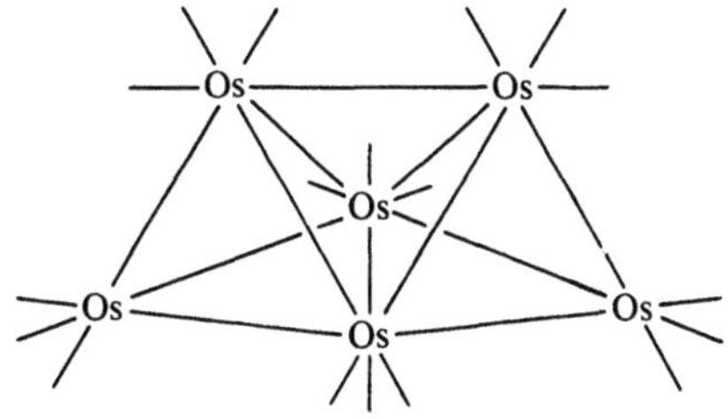

12

A considerable number of clusters with more than six metal atoms has been prepared. There is no structural systematics for these yet, but in one case of a Rh_{13}-cluster (*12*) it has been shown that the packing of metal atoms in a big cluster can be the same as that of a closely packed array in the pure metal. In the following structural drawings all ligands are omitted to allow a clear view of the metal atom arrangement.

The seven-atom clusters, $Os_7(CO)_{21}$ (*151, 284*) and $Rh_7(CO)_{16}J^{2-}$ (*19*), adopt the structure [*13*] of a monocapped octahedron. Among the eight-atom clusters the cube-like Cu_8-compounds were known and have been confirmed by further examples (*70, 206*). In addition to this, the structures of a bicapped octahedron or a bicapped trigonal prism have been suggested for $Os_8(CO)_{23}$ and $Os_8C(CO)_{21}$ (*151*), and the unusual structure [*14*] of a monocapped and mono-edge-bridged trigonal prism has been proved for $Rh_8C(CO)_{19}$ (*14, 23*).

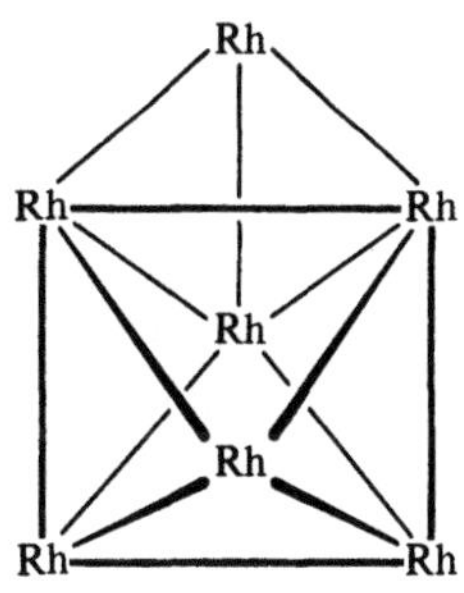

13

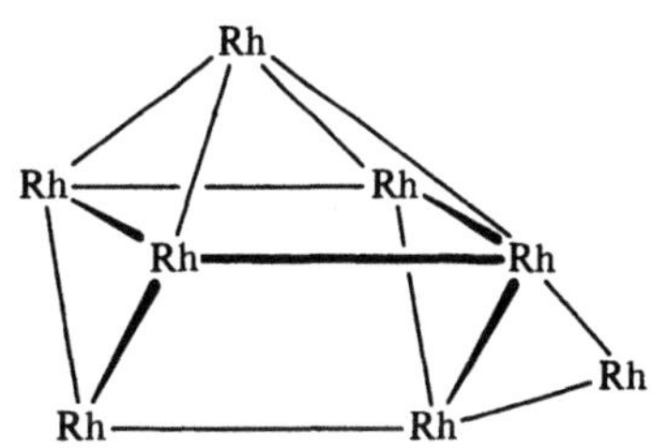

14

Several new polynuclear gold clusters have been obtained (*37, 38, 77*), and the structures of an Au_9 and another Au_{11} cluster have been determined (*37, 39*). Their basic atom arrangements [*15*] and [*16*] can be derived from centered icosahedra by removing certain vertices.

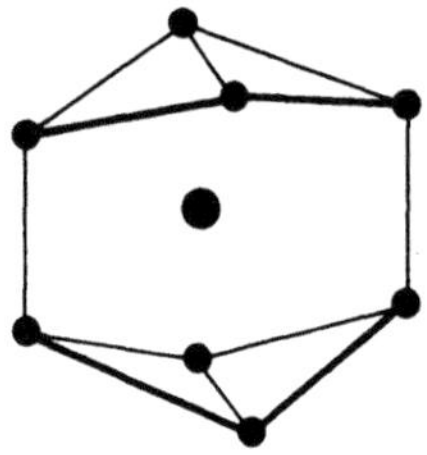

15

● = Au

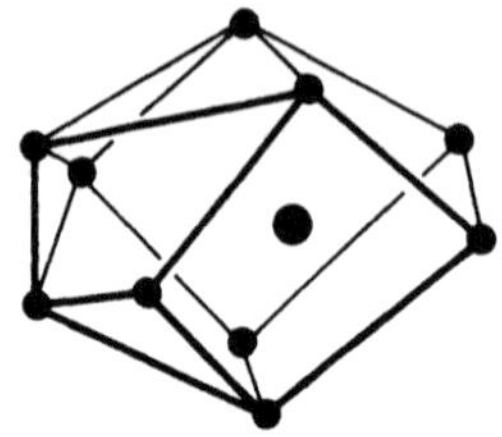

16

The Rh_{13} cluster $H_3Rh_{13}(CO)_{24}^{2-}$ (*12*) has the very regular atom arrangement *17*, which is identical to that in a hexagonally closely packed metal. This is not the case for $Rh_{15}C_2(CO)_{28}^{-}$ (*14*), the structure [*18*] of which can be regarded as a centered tetracapped pentagonal prism. However, in this cluster, too, the central rhodium atom has the metallic coordination number 12.

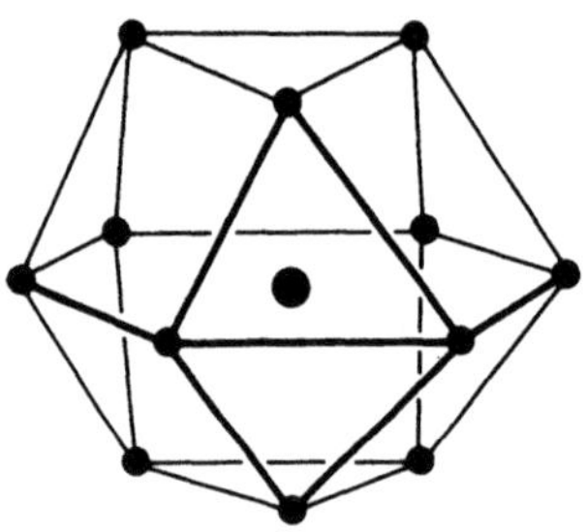

17

● = Rh

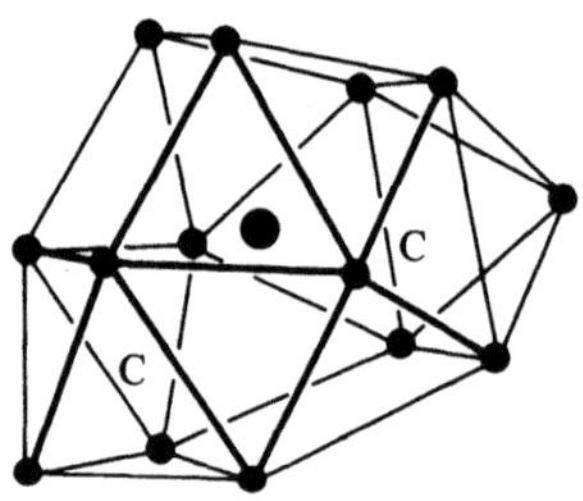

18

A most unusual series of clusters has been obtained and structurally characterized as belonging to the general composition $[Pt_3(CO)_6]_n^{2-}$ with n = 2, 3, 4, 5 (*72*). All of these clusters are composed of $Pt_3(CO)_6$ units stacked on top of each other. [*11*] is the structure for $n = 2$, [*19*] for $n = 3$, and [*20*] for $n = 5$. The prediction seems justified that the polymeric $Pt(CO)_2$ will have the same chimney-like structure where the two negative charges can be neglected due to the high value of *n*.

Clusters with more than 15 transition metal atoms have not been fully characterized as yet.

Unusual Compositions. Three series of compounds deserve special consideration because they have developed from laboratory curiosities into classes of their own. These are the heterometallic clusters, the carbide clusters, and the tetranuclear cubane clusters.

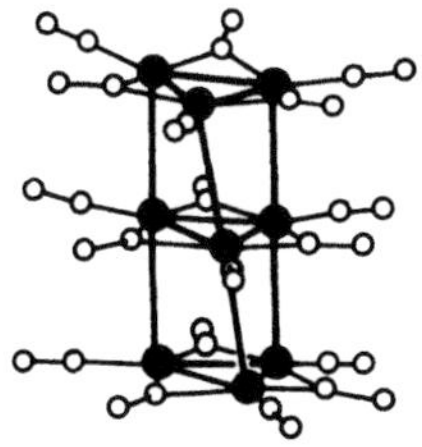

19

● = Pt

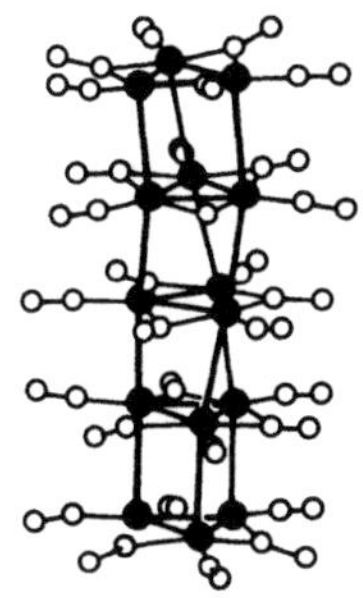

20

A considerable number of heteronuclear clusters have been prepared and investigated during the last five years. Listed with the metal atom later in the periodic table, there are hetero clusters containing iron (*106, 215*), ruthenium (*7, 57, 175, 191, 234, 248, 249, 293*), osmium (*248, 249*), cobalt (*106, 107, 212, 216, 366, 368, 394, 403*), rhodium (*14, 97, 98, 212, 228, 278*), iridium (*21, 278*), nickel (*20, 87, 213, 333*), platinum (*15, 16, 62, 63, 64, 134, 285, 286, 295*), copper (*8, 91*), silver and gold (*262, 265, 272*).

Most of these complexes adopt structures of the familiar types. Only in a few cases have unusual structures (*91, 249, 333*) been observed as well as compositions or structures not obtainable with the isoelectronic homonuclear compounds (*97, 98*).

Five years ago only two carbido metal complexes were known (*241*), $Fe_2C(CO)_{15}$ and $Ru_6C(CO)_{17}$, and their mode of formation and the source of the carbon atom were obscure. Since then carbon atoms have been found in the centers of several metal atom polyhedra; and it has been shown that the carbon atoms stem either from solvent molecules (*13, 22*) or from carbonyl ligands of the starting materials (*149, 151*), or may even be introduced in a semi-systematic manner via the C–Cl function of methinyl tricobalt enneacarbonyls (*14*).

There are three iron carbide clusters now, $Fe_5C(CO)_{15}$ (*108, 367*) and $Fe_5C(CO)_{14}^{2-}$ (*214*) with the tetragonal pyramidal structure *9*, and $Fe_6C(CO)_{16}^{2-}$ (*101, 102, 367*) with the regular octahedral structure. Of ruthenium and osmium both the analogous compounds $Ru_5C(CO)_{15}$ (*152*) and $Os_5C(CO)_{15}$ (*149, 151, 152*) have been obtained, together with the known $Ru_6C(CO)_{17}$ and the new compound $Os_8C(CO)_{21}$ (*151*), for which the unusual structure of a bicapped octahedron or a bicapped trigonal prism has been suggested. The only carbido cluster containing only cobalt is $Co_6C(CO)_{15}^{2-}$ (*14*), which is similar to the rhodium-containing compounds $Rh_4Co_2C(CO)_{15}^{2-}$ (*14*) and $Rh_6C(CO)_{15}^{2-}$ (*13, 22*), which was the first example of a trigonal prismatic cluster. Rhodium has also yielded the unique carbido clusters $Rh_8C(CO)_{19}$ (*14, 23*) and $Rh_{15}C_2(CO)_{28}^{-}$ (*14*) with the structures *14* and *18*. Further carbido clusters are to be expected, since the introduction of carbon atoms seems to be more than a side reaction.

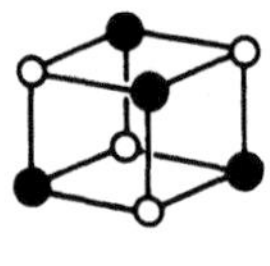

21

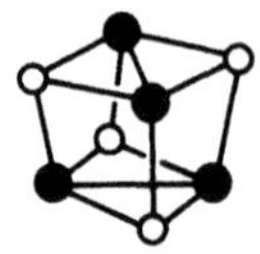

22

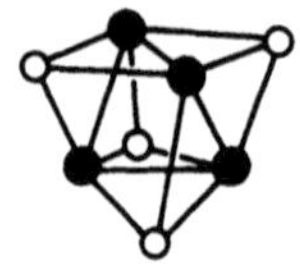

23

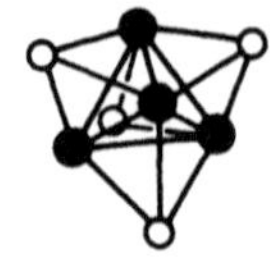

24

The stability of the many cubane-like M_4E_4 complexes results from the fact that they can be held together by metal-metal bonds as well as by the clamp effect of the triply-bridging ligands E. This allows for all degrees of metal-metal bonding within the metal atom tetrahedron that constitutes one half of the cubane skeleton. The range of metal-metal interactions reaches from completely nonbonding, as in a number of main group and transition metal cubanes (*18, 204, 361*), to completely bonding (six M–M bonds in valence bond notation) as in $[CpFe(CO)]_4$ (*303*), $[(NO)FeS]_4$ (*180*) and $[(CO)_3Re\text{-}In\text{-}Re(CO)_5]_4$ (*202*). The intermediate situation of two metal-metal bonds is realized in $[CpFeS]_4$ (*381*) and $[CpCoP]_4$ (*362*), whereas four metal-metal bonds occur in $[\{(CF_3)_2C_2S_2\}FeS]_4^{2-}$ (*41*) and $[(NO)Co(NCMe_3)]_4$ (*181*). The metal and bridging ligand atom arrangements with 0, 2, 4 and 6 metal-metal bonds are depicted in *21, 22, 23,* and *24*.

Bonding concepts for cubanes, as discussed in Chapter 2.3., predict that the metal polyhedra in the cubanes will contract upon oxidation because of increased metal-metal interaction and vice versa. This has been verified for $[CpFeS]_4^{n+}$ (*381*), $[CpFe(CO)]_4^+$ (*167, 177*), and $[CpCoS]_4^+$ (*361*). Furthermore, the ease of oxidation and reduction of several cubane-type clusters (*166, 167, 361, 381*), and the delocalization of electrons in the charged species (*48, 176, 177, 401*) is noticeable. This, together with the prefered formation of iron-sulfur clusters, (*381*), is borne out by the fact that Nature uses iron-sulfur proteins for redox reactions (*207*).

Fluxionality. The fluxional behavior of unsaturated organic ligands in mono- and oligonuclear organometallic complexes was well documented (*109a*) as a result of extensive ^1H–NMR spectroscopic investigations. The advent of pulse Fourier transform ^{13}C–NMR spectroscopy has extended these investigations to movements of CO groups and similar ligands at metal atoms in, and around all metal atoms of, clusters. In fact, the major breakthrough in ^{13}C–NMR spectroscopy of carbonyl ligands, namely the introduction of $Cr(acac)_3$ as a relaxation agent, was achieved together with the investigation of the very fluxional molecule $Fe_3(CO)_{12}$ (*186*).

It has been shown that most of the binary metal carbonyl clusters are fluxional. These include $Fe_3(CO)_{12}$ (*9, 111, 113, 175, 186*), $RuFe_2(CO)_{12}$ and $Ru_3(CO)_{12}$ (*9, 175*), which show complete CO scrambling at room temperature. $Os_3(CO)_{12}$ is rigid up to 70° (*9, 175*), whereas $Os_6(CO)_{18}$ shows localized scrambling at room temperature (*148*). All CO groups move in $Co_4(CO)_{12}$ at room temperature (*103, 162*), but in $Rh_4(CO)_{12}$ only above 60° (*110, 112, 164*). The reduced scrambling tendency for the heavier metal clusters is well demonstrated by $RhCo_3(CO)_{12}$ (*228*), in which at 10° all CO groups except at the $Rh(CO)_2$ unit are mobile. Of the several other rhodium carbonyls $Rh_6(CO)_{16}$ (*203*) and $Rh_{12}(CO)_{30}^{2-}$ (*90*) are rigid, whereas $Rh_7(CO)_{16}^{3-}$ (*203*) shows intermediate and $Rh_6(CO)_{15}^{2-}$ (*203*) shows rapid fluctuation. Mixed metal carbonyls (*175, 228*) and cyclopentadienyl metal carbonyls (*212*) provide the proof that CO scrambling goes around the whole cluster. Lewis base substituted carbonyl clusters of iron (*114*), cobalt (*103, 288*), and iridium (*80*) behave similar to the parent carbonyls, as do isonitrile clusters (*122, 192*) in which isonitrile and carbonyl ligands scramble. Ligand influence on CO mobility is obvious in $Ru_3(CO)_{10}(NO)_2$ (*175*) where only the CO groups on the ruthenium atom not bonded to the NO ligands show localized scrambling.

Hydride scrambling is a common process in hydride carbonyl clusters, and it has been found that the energies of activation for this process are similar to those for the movement of hydrogen atoms in plain metals (*233*). The temperature dependence of the NMR spectra has been used to gain structural information on $H_2FeRu_3(CO)_{13}$ (*293*) and on derivatives of $H_4Ru_4(CO)_{12}$ (*250, 254*). The relative mobilities of H and CO ligands have been compared in $HFe_3(CO)_{11}^-$ (*175*).

Acetylenic ligands in carbonyl clusters may be rigid as in $Co_4(CO)_{10}$ (RC=CR) (*10, 162*), $Fe_3(CO)_8(Ph_2C_2)_2$ (*10*), and $HM_3(CO)_9$ (C≡CR) (*9*) (M = Ru, Os), or able to rotate above a triangle of metal atoms as in $Cp_3M_3(CO)$ (ArC≡CAr) (M = Rh, Ir) (*188, 400*) and $H_2M_3(CO)_9$ (RC≡CR) (M = Ru, Os) (*163*). The latter complexes were also the subject of a detailed study (*163*) which showed four fluxional processes to be present: (1) hydrogen exchange, (2) local CO scrambling, (3) total CO scrambling, (4) rotation of the acetylene ligand.

Triangular osmium complexes show many unusual kinds of ligand bonding (see Chapter 3.4.), and this behavior extends to their fluxional properties. Thus, the clusters $HOs_3(CO)_{10}$ (CH=CHR) show a ligand oscillation of the type [*25a*] ⇋ [*25b*] (*358*), and the ortho-metalated phenyl clusters $Os_3(C_6H_4)(EMe_2)_2(CO)_7$ *26* (E = P, As) (*1*) exhibit a rotation of the C_6H_4 group together with CO scrambling. On the contrary, in the clusters $H_2Os_3(CO)_9$ C=CH_2 (*224*) and H_2Os_3(CO), (CH–PMe_2) (*132*) H-atom scrambling occurs without movement of the unusual triply briding ligands C=CH_2 and CH–PMe_2.

Ligand fluxionality and catalytic activity are interrelated, and considerable insight into the reactivity of clusters is to be expected from further studies of this type.

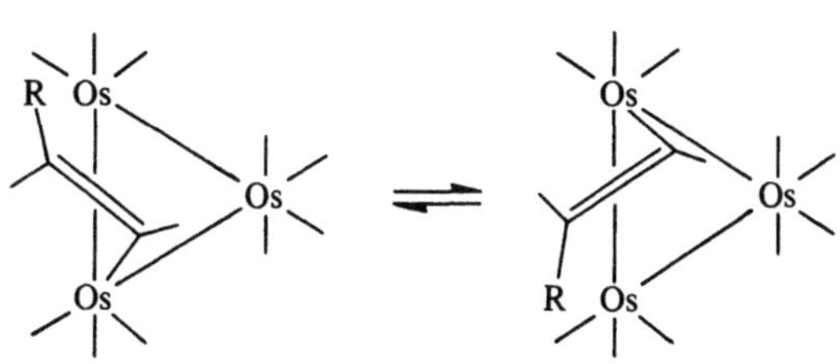

25a *25b*

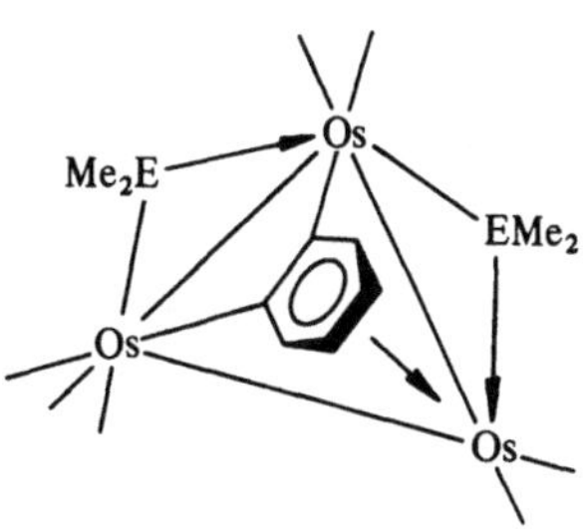

26

2.3. Bonding

With metal clusters it is even harder than in other fields of inorganic chemistry to substantiate theoretical results by energy measurements. Only two such measurements have come to the attention of the author – the photoelectron spectrum of $[CpFe(CO)]_4$ (*370*) and bond energy determinations in $Co_3(CO)_9CX$-compounds (*187*). However, a considerable number of papers deal with metal-metal bonding in, and the symmetry properties of, clusters as related to their stoichiometry and their electron count. These studies have confirmed the wide applicability of the simple 18-electron rule in predicting metal-metal bonds and structures, but they have also led to an understanding of the limits of this rule for clusters with more than four metal atoms.

Metal-metal bonding in ligand bridged oligonuclear complexes has been treated in terms of symmetry properties of metal a.o.'s (*281, 282*) and of a parameter free MO model (*198, 373*). The most important result of the later study is that in these complexes HOMO's and LUMO's are composed predominantly of metal d orbitals and represent metal-metal σ-interactions. Therefore a change of the number of electrons in a cluster with up to four metal atoms will affect primarily the metal-metal bonding. Depending on the occupancy of the HOMO's this change can be toward more bonding or more antibonding metal-metal interactions.

The HOMO/LUMO nature of the metal-metal bonds in clusters has been substantiated by the photoelectron spectrum of $[CpFe(CO)]_4$ (*370*) and by magnetic field Mössbauer spectra of $[S_2C_2(CF_3)_2]_4Fe_4S_4$ (*176*) and $[CpFe(CO)]_4$ (*177*) and oxidized and reduced species thereof. The effect of addition and removal of electrons by oxiation/reduction reactions on metal-metal bonds (*178, 373*) in all cases supports the proposed energy levels of metal-metal interaction. In one case the experimental "location" of an electron in the HOMO of a cluster was possible. This was achieved by a single crystal EPR study of paramagnetic $ECo_3(CO)_9$ (E = S, Se) [*27*] diluted in the diamagnetic host $EFeCo_2(CO)_9$ [*28*] (*366, 368*). The antibonding nature of the HOMO is demonstrated by the lengthening of the cobalt-cobalt bonds in [*27*] relative to those in [*28*] which contains one electron less. The EPR results place this antibonding electron in a σ–MO of a_2 representation in accord with a symmetry-based MO treatment (*368*).

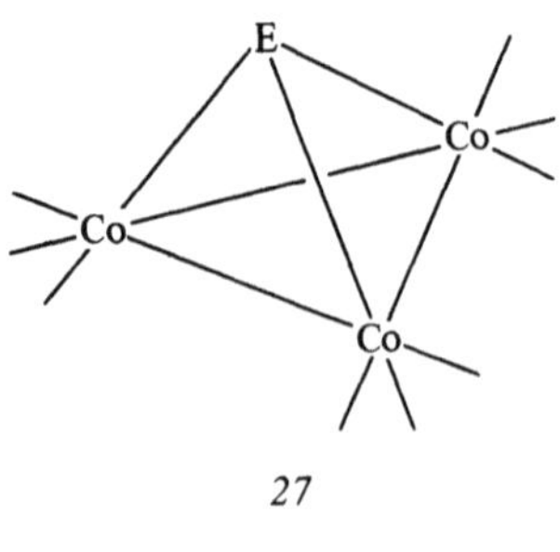

27

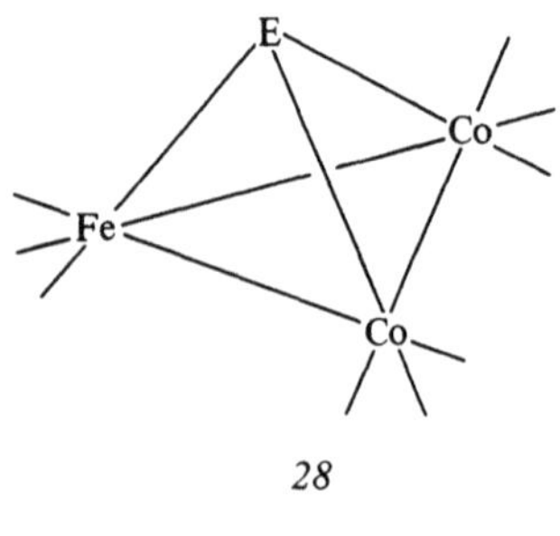

28

Tetrametallic cubane clusters are the ones most frequently dealt with in MO treatments with respect to the degree of metal-metal bonding. Qualitative MO arguments have rationalized the geometries adopted by $[CpFe(CO)]_4$ (*375*), $[(NO)Co(\mu_3 - NCMe_3)]_4$ (*181*), $[CpCoS]_4$ (*361*), $[CpCoP]_4$ (*362*), and the deformations of these clusters upon oxidation and reduction. More elaborate treatments have been used for the iron-sulfur cubanes. The parameter-free model has been applied to $[CpFeS]_4$, $[(NO)FeS]_4$, and $[(RS)FeS]_4^{2-}$ (*180*), showing considerable differences in orbital arrangements between these three compounds and making it understandable why the first two of them are not suited as models for the four iron-four sulfur redox proteins. And finally, the X_α–SCF–MO technique has been applied to $[(MeS)FeS]_4^{2-}$ (*401*) allowing detailed conclusions but confirming the results of the simpler MO models as to the HOMO/LUMO character of the metal-metal interactions.

Clusters with up to four metal atoms generally obey the 18-electron rule for each metal atom. This is no longer the case for six-atom clusters. For instance, in an octahedral cluster a total of 84 electrons (provided by the metal atoms and the ligands) would be predicted for a stable configuration, counting each edge of the octahedron as a metal-metal bond. The 84-electron count for an octahedral cluster is realized in only one known case, namely $[HCu(PPh_3)]_6$ (*42, 92*), and almost all other cases, as listed in ref. (*20*), have 86 electrons.

The electron-counting scheme which allows to understand this situation and which also works for the simpler clusters was developed from the MO theory for boron hydride polyhedra (*389–393*). It rests on the fact that a regular triangular faced polyhedron of n boron atoms in the molecule $B_nH_n^{2-}$ requires n + 1 bonding skeletal electron pairs, e.g. 7 bonding electron pairs for a $B_6H_6^{2-}$ octahedron. Changing the number of bonding electron pairs will change the shape of the polyhedron: an increase will result in a more open (nido or arachno) structure (*393*), a decrease will result in a more closed (monocapped or bicapped) strucutre (*151, 284*). Thus, for metal atoms the following cluster structures are predicted (n is the number of metal atoms in the cluster):

number of skeletal bonding electron pairs	shape of the cluster
$n + 3$	arachno (2 vertices unoccupied in a $(n + 2)$-polyhedron)
$n + 2$	nido (1 vertex unoccupied in a $(n + 1)$-polyhedron)
$n + 1$	regular n-polyhedron
n	monocapped $(n - 1)$-polyhedron
$n - 1$	bicapped $(n - 2)$-polyhedron

In order to make transition metal clusters comparable to borane polyhedra, the following counting scheme is adopted: (1) each metal atom uses one orbital for each of its ligands; (2) each metal atom uses three orbitals for skeletal (cluster) bonding; (3) each metal atom keeps its remaining orbitals occupied with nonbonding electron pairs. Applied to $Rh_6(CO)_{16}$ with a Rh_6-octahedron one has to reserve $6 \cdot 3 = 18$ a.o.'s first for skeletal bonding. Of the 86 available electrons, 72 are placed in the $6 \cdot 6 = 36$ remaining nonbonding and metal-ligand bonding orbitals. This leaves 14 electrons, i.e. seven bonding skeletal electron pairs as required for the octahedral six-atom polyhedron. Thus, this skeletal electron-counting method is superior to the 18-electron rule for octahedral clusters, but both counting schemes agree in requiring 60 electrons for tetrahedral and 54 electrons for triangular clusters.

A justification for the skeletal electron-counting method has resulted from a Wolfsberg-Helmholz-MO calculation on the very symmetrical anion $Co_6(CO)_{14}^{4-}$ (*294*). This calculation again revealed that in clusters HOMO's and LUMO's are composed predominantly of metal d-orbitals. It turned out, however, that the octahedral arrangement of metal atoms does not go along with seven strongly bonding skeletal MO's but rather with 11 strongly antibonding MO's which remain empty when 86 electrons are present.

The skeletal bonding-electron pair approach works for octahedral 86-electron clusters and for some bigger clusters. It can also explain why $Os_6(CO)_{18}$ with 84 electrons adopts the structure of a monocapped trigonal bipyramid (*151, 284*). It does not explain, however, why $[HCu(PPh_3)]_6$ (*42, 92*) with 84 electrons is octahedral as well as $Ir_2Cu_4(PPh_3)_2(C{\equiv}CPh)_4$ (*91*) with 90 electrons. Neither does it explain why $Rh_6C(CO)_{15}^{2-}$ (*22*) with 90 electrons is trigonal prismatic like $Pt_6(CO)_{12}^{2-}$ (*72*) with 86 electrons, which differs markedly from its octahedral analog $Ni_6(CO)_{12}^{2-}$ (*71*).

Of the clusters with more than six metal atoms, the structures [*13*] of $Os_7(CO)_{21}$ (*151, 284*) and $Rh_7(CO)_{16}J^{2-}$ (*19*) can be rationalized in terms of skeletal electron-pair counting. For the clusters with six (*38*), nine (*37*), and eleven (*39*) gold atoms the rules have to be modified in order to agree with the observed structures, the main assumption being that one orbital per gold atom remains empty (*38*). For clusters of 8 to 15 metal atoms there is no correlation between structures and electron counts yet.

Metal-ligand bonding is assumed to be separable from metal-metal bonding in the simple electron-counting schemes outlined above. The MO calculations give support to this by showing that MO's associated with ligand bonding are normally lowest in energy and that changes in the number of electrons affect the metal-metal bonding primarily. Nevertheless, reactivity of clusters is mostly a matter of metal-ligand bonding. Therefore the attachment of unusual ligands to clusters is widely sought for. The known unusual bonding situations, especially between unsaturated organic groups and several metal atoms, have not been treated in MO terms, however. They will be discussed therefore together with the corresponding metal clusters in Chapter 3.

2.4. Syntheses

No systematic synthesis of any cluster has been devised as yet. The only generalizable statement in cluster preparations is that clusters may be formed when a coordinatively unsaturated metal species is generated in the absence of donor ligands (*231, 241*). Consequently, clusters are often formed as unexpected products in reactions where the formation of unsaturated metal species is imaginable but not obvious. Whereas these obscure formations of clusters cannot be treated in this chapter, some preparative procedures used more often deserve comment.

One of them is the reduction of simple metal salts in the presence of CO or similar ligands, the most general starting reaction in organometallic chemistry. The method has been used to prepare all basic metal clusters, and during the last five years improved procedures have been described for $Ru_3(CO)_{12}$ and $Os_3(CO)_{12}$ (*121, 227, 245*) $Rh_4(CO)_{12}$ (*81, 279*) and $Ir_4(CO)_{12}$ (*275, 369*). Also for hydride clusters reduction is an applicable synthetic pathway (*233*), often achieved by disproportionation reactions under the influence of hard bases (*17, 24*) or by hydride incorporation from main group metallic hydrides (*42, 92, 298, 299*). Hydride reduction without incorporation of the H-ligand is used in the synthesis of the new Au_6^- (*38*) and Au_9^- (*37*) clusters, whereas for the triangular Pd-clusters $Pd_3(CO)_3(PPh_3)_3$ ($n = 3, 4$) CO is a sufficiently strong reducing agent for the palladium salts used as starting materials (*267*). Quite often anionic clusters are the result of reduction reactions, especially in hydride chemistry (*233*), but also with rhodium (*14, 277*), nickel (*71, 274*), and platinum carbonyls (*72*). Clusters themselves may be the subject of reduction or disproportionation reactions. These cases are dealt with in Chapter 2.5.

The second most important procedure to generate unsaturated species is pyrolysis or photolysis of simple complexes best known from the conversion of $Fe(CO)_5$ into $Fe_2(CO)_9$ and $Fe_3(CO)_{12}$. Simple reactions of this kind reported recently include the formation of $Cp_3Co_3(CO)_3$ and $Cp_4Co_4(CO)_2$ from $CpCo(CO)_2$ (*386, 387*), the formation of $Cp_3Co_3(NO)_2$ from $[CpCo(NO)]_2$ (*301*), and the formation of $Cp_3Ru_3(CO)_2(SR)_3$ from $Cp(CO)_2Ru{-}SR$ (*252*). Reactions at ligands of simple complexes are involved in the formation of $Me_3SiO{-}CCo_3(CO)_9$ from $Me_3Si{-}Co(CO)_4$ (*220*), the formation of $Co_3(CO)_9CX$ from $Co_2(CO)_6$ ($RC{\equiv}CR$) (*140*), and the formation of $[(CO)_3M{-}GeMe_2]_3$ from $(CO)_4M(GeMe_3)_2$ with M = Ru, Os (*253*). Several "big" clusters of ruthenium and osmium have been obtained by pyrolysis of the small clusters $Ru_3(CO)_{12}$ and $Os_3(CO)_{12}$ (*149–151*). Somewhat in between metal cluster synthesis and metal cluster reactivity fall the pyrolysis reactions of phosphine substituted $Ru_3(CO)_{12}$ and $Os_3(CO)_{12}$ derivatives, in that they proceed at clusters but mainly involve reactions at the ligands. Thus, the pyrolysis of a mixture of $Os_3(CO)_{12-n}(PPh_3)_n$ ($n = 1, 2, 3$) yields six $Os_3(CO)_x$ clusters in which phenyl groups of the phosphine ligands have been introduced as newly bound ligands (*50, 51*). Similar reactions have been observed with aryl-phosphine-substituted $Ru_3(CO)_{12}$ derivatives (*65, 66, 128*), and even methyl groups of methyl phosphine ligands undergo pyrolytic formation of Os–C bonds in certain osmium carbonyls (*132*). Here the formation of coordinatively unsaturated metal centers and coordinatively unsaturated ligands work hand-in-hand.

A similar situation occurs with the third most important method of cluster synthesis, namely the use of unusual ligands which bear the ability to bridge several metal atoms. These ligands may themselves be only short-lived intermediates, but find a suitable bonding situation over the face or edge of a cluster polyhedron. Well-known examples of this type are the nitrene ligands RN bridging three metal atoms and resulting from such sources as organic azides (*1, 255*), nitro compounds (*26*), or sulfur diimides (*181, 236*). Recently the analogous phosphinidene ligands RP have been introduced as triply- and quadruply-bridging groups resulting from compounds like $RPCl_2$ (*276, 334, 385*) and RPH_2 (*374*). The carbyne ligand RC in methinyltricobalt enneacarbonyls normally results from a compound $RCCl_3$ (*347*), but may also stem from a cobalt-bound acetylene (*140*), or even from a carbyne ligand bonded to another metal (*173*). The most common unusual triply-bridging ligand is the sulfur atom (*381*), and many sources of sulfur atoms are known, demonstrating the pronounced ability of sulfur to assist the aggregation of metal atoms. They include elemental sulfur (*180, 361*) and sulfur-releasing compounds, such as episulfides (*35*) and disulfides (*82, 196, 197*), but also compounds with sulfur-nitrogen (*178*) sulfur-tin (*384*) or sulfur-phosphorus bonds (*302*), and even the sulfite ion (*335*). A similar metal-aggregating ability is observed for acetylene, and in recent years this has been further demonstrated by the formation of acetylene-bridged clusters of iron (*311*), nickel (*120*), and rhodium (*138, 139, 188, 189, 242, 326*).

Less common ligands used to induce cluster formation include azo compounds (*24, 25*), sulfur dioxide (*309, 310*), and main group elements such as boron (*339a*), aluminium (*341*), gallium (*201*), indium (*202*), phosphorus, and arsenic (*135*), which are introduced either in elemental form or as halides. A promising development in cluster chemistry seems to be the use of ligands containing metal atoms themselves. This has been shown so far by reactions between metal-substituted acetylenes and simple complexes which yielded acetylene-bridged clusters containing Fe_3 (*6, 403*), Fe_2Ru (*7, 57*), $FeCo_2$ (*403*) and Cu_4Ir_2 (*8*) units.

The new field of heterometallic clusters was also opened by reactions which introduce new metal atoms into existing (mostly oligonuclear) complexes. These reactions between different metal complexes may be called the fourth general synthetic procedure in cluster chemistry. The simplest application is the combination of two neutral compounds, which was successfully used in the reactions of several oligonuclear metal carbonyls with platinum phosphine complexes PtL_4 (*16, 62, 63, 64, 286*) or cyclopentadienyl metal carbonyls (*213*). A semi-systematic cluster synthesis is provided by the reactions of halogen-containing metal complexes with anionic metal-carbonyls. Several mixed metal carbonyls containing cobalt and/or rhodium and iridium have been obtained this way (*14, 208, 278*) as well as some nickel-cobalt carbonyls (*87*). Platinum is again introduced into heterometallic clusters this way via L_2PtCl_2 (*52, 134, 174*). And finally, the reaction between neutral and anionic metal carbonyls leads to anionic or hydride clusters, as has been shown using the anions

$M(CO)_5^-$ (M = Mn, Re) (*102, 249*), $M(CO)_3Cp^-$ (M = Mo, W) (*215*), and $M_2(CO)_{10}^{2-}$ (M = Cr, Mo, W) (*333*). Further clusters from such reactions are to be expected.

With more and more clusters becoming available, the synthetic chemistry of clusters turns more to the use of clusters as starting materials. Thus, there is an extensive literature on CO substitutions in metal carbonyl clusters and on the organic chemistry of methinyltricobalt enneacarbonyls. Reactions of this type are dealt with in part in Chapter 2.5. and in detail in Chapter 3.

2.5. Reactions

It is well known that many clusters, especially those of the first-row transition metals, undergo degradation upon reaction with compounds that possess nucleophilic properties, i.e. CO, phosphines and related Lewis bases, unsaturated organic compounds, and reducing agents. This is in agreement with the results of the theoretical calculations which state that metal-ligand bonding liberates more energy than metal-metal bonding. Reactions of this type have been dealt with in detail (*231, 241*) and will not be treated here. This chapter is confined to some common aspects of cluster reactions which lead to products that are clusters again. Attention will be given to cluster-rearrangements, substitution reactions with Lewis bases and unsaturated organic compounds, some redox and oxadative addition reactions, and some aspects of hydrogenation and homogeneous catalysis. In comparing previous reviews (*86, 231, 241*) one notes that cluster reactivity is the field that has been most developed.

Rearrangements of clusters, i.e. changes of cluster shape and increase and decrease of the number of cluster metal atoms, have already been mentioned with pyrolysis reactions and heterometallic cluster synthesis in chapter 2.4. Furthermore, cluster rearrangements can occur under conditions which are similar to those used to form simple clusters, e.g. simple redox reactions interconvert four to fifteen atom rhodium clusters (*12, 14, 280*). Hard-base-induced disproportionation reactions lead to many atom clusters of rhenium (*17*), ruthenium and osmium (*233*), iron (*108*), rhodium (*22, 88, 277*), and iridium (*28*). And the interaction of metal carbonyl anions and clusters produces bigger clusters of iron (*102, 367*), ruthenium, and osmium (*249*).

The most common substitution reactions of clusters are the replacements of CO groups by phosphines and arsines. Under controlled conditions, to avoid subsequent reactions and cluster degradations, these can be achieved for clusters of iron (*83, 84, 108, 116, 325, 331*), ruthenium (*58, 65, 66, 230, 307, 318, 327, 377*), osmium (*50, 118, 128, 132*), cobalt (*119, 155, 268, 289, 363*), rhodium (*45, 221, 396*), iridium (*80, 147*), and of mixed metal carbonyls (*106, 107*). Kinetic investigations of these reactions have shown mostly regular substitution behavior, i.e. first order reactions initiated by CO dissociation (*78, 238, 307, 331*). However, for $Ru_3(CO)_{12}$ several reaction paths were observed, including one with a mononuclear intermediate which trimerizes again (*319–321*); and in the case of $Ir_4(CO)_{12}$ cooperative substitution was found, i.e. the ease of CO-substitution is increased with increasing number of phosphine ligands present (*238*). Isonitriles with their better acceptor properties replace more CO groups than phosphines and lead to less degradation of clusters (*83, 192, 304*). Substitution of CO groups by halide and other ions has been used so far only for rhodium clusters (*89*). The most common subsequent reaction of cluster substitution that does not involve degradation is the ortho-metalation of cluster-bound phosphine and arsine ligands. This will be treated as an oxidative addition reaction (see below). Additions rather than substitutions are also the reactions of $H_2Os_3(CO)_{10}$ with Lewis bases, which lead to the clusters $H_2Os_3(CO)_{10}L$ by loss of the Os–Os double bond (*124, 357*).

Like the Lewis base substitutions, the reactions of metal clusters with unsaturated organic compounds are confined to the core area of clusters in the periodic table, i.e. to the clusters of the iron and cobalt groups. However, the different hydrocarbons with double and triple bonds offer many more reaction paths because of cluster-induced transformations of these hydrocarbons. This is the basis of catalytic processes such as hydrogenation, isomerization, and oligomerization, which can also be achieved with metal clusters (see below). The hope for the isolation of possible reaction intermediates and for information on possible pathways of catalytic reactions has stimulated many reactions discussed here during the last five years. The present results show that there are some clusters that have just been substituted by the unchanged hydrocarbon, but that in many cases the hydrocarbon has undergone an organic reaction before being trapped as a ligand. The structural drawings necessary to clarify the following discussion may be found in the chapters on the corresponding metals.

Unchanged polyenes have been found to form complexes with methinyltricobalt enneacarbonyls (*53, 159, 160, 225, 330*) and with cobalt and rhodium carbonyl clusters (*247*). Arenes substitute three carbonyl groups in $Co_4(CO)_{12}$ (*43, 47, 240, 363*) and in $Co_3(CO)_9CX$ (*53, 137, 328*), whereas azulene replaces five CO groups in $Ru_3(CO)_{12}$ (*96, 100*). Mono-acetylenes are bound, though not in a simple fashion, to iron (*108*), ruthenium (*183, 229*), osmium (*371*), cobalt (*106, 363*), and rhodium (*45*) clusters. $Ru_3(CO)_{12}$ frequently induces organic reactions of cyclic olefins prior to cluster complex formation. Among these are dehydrogenation (*36, 59, 74, 75, 283*), hydride transfer (*32, 33, 68*), and ring closure, the most prominent example of which is the formation of pentalene-Ru_3 complexes from cyclooctatetraenes (*33, 209, 210*). $Os_3(CO)_{12}$, on the other hand, has a pronounced tendency to form Os_3 complexes containing the osmacyclopentadiene ring composed formally of two acetylenes and one osmium atom. This has been amply demonstrated for reactions with diphenylacetylene (*168–171, 184, 371*), but also with other acetylenes (*182, 223, 379*).

The most important reaction of unsaturated organic substrates with respect to homogeneous catalysis is the oxidative addition of C–H bonds to ruthenium and osmium carbonyl clusters. This reaction which was first discovered as the so-called ortho-metalation of phosphine ligands in mononuclear complexes was subsequently also found for cluster-bound phosphine ligands. First results with triphenylphosphine derivatives of $Os_3(CO)_{12}$ (*50, 51, 179*) and $Ru_3(CO)_{12}$ (*65*) which led to a host of complexes with ortho-phenylene groups in unusual bonding situations together with metal-hydrogen bonds, were subsequently extended to aryl phosphite (*66*) and methylphenyl-phosphine complexes (*128, 129*) with similar results. And finally, it was found that even alkyl phosphines (*132*), phenyl acetylenes (*168, 169*), and pyridines (*407*) undergo oxidative additions of C–H bonds to osmium carbonyl clusters. With ruthenium carbonyl clusters several open and cyclic olefins give similar reactions which lead to hydride clusters containing the partially dehydrogenated olefins bound to the metal atoms via σ-, π-, and π-allyl interactions. The olefins used are butadienes (*185*), hexadienes (*165*), cyclopentadiene (*217*), cyclooctadienes (*74*), bicyclooctadiene (*146*), and cyclododecatriene (*59, 117*). And finally, the C–H bonds of ethylenes and simple acetylenes bind with ruthenium and osmium carbonyls unter appropriate conditions. $Ru_3(CO)_{12}$ and RC≡CH react to give $HRu_3(CO)_9$ (C≡CR) (*190, 336*). And the simplest reaction between ethylene and $M_3(CO)_{12}$ (M = Ru, Os) produces $H_2M_3(CO)_9$ ($C{=}CH_2$) (*127, 131, 133*). This reaction has been studied in some detail, and possible intermediates have been isolated using ethylenes, acetylenes, $Os_3(CO)_{12}$, and $H_2Os_3(CO)_{10}$ (*125, 126, 222, 239*).

It has been found in the meantime that reaction (1) is generalizable (*132*), and that oxidative additions of this type occur for such widely differing substrates H_2Y as ethylene, benzene (*130*), cyclic olefins, alkyl and aryl phosphines, aniline (*337, 406*), and H_2S (*130*), all of which give the same product structure with a triply-bridging Y ligand. The stability of these third-row transition metal clusters has still prevented catalytic reactions of these species, but it is likely that similar ones are involved in olefin and acetylene reactions catalyzed by other metal complexes.

$$Os_3(CO)_{12} + H_2Y \rightarrow 3\,CO + H_2Os_3(CO)_9Y \qquad (1)$$

The simplest reaction which is made possible by the metal-metal bonded framework of the clusters is the addition and removal of electrons. Due to the HOMO/LUMO nature of the metal-metal bonding MO's, the transfer of electrons to and from clusters requires relatively small structural changes because metal-ligand bonding is not affected. Electrochemical investigations on a number of clusters have confirmed this (*292, 315*), and a number of cluster cations and anions with uneven numbers of electrons have been isolated. Most redox reactions have been achieved with cubane clusters such as $[CpFe(CO)]_4$ (*167*), $[CpFeS]_4$ (*166*), $Fe_4S_4[S_2C_2(CF_3)_2]_4$ (*176*), and $[CpCoS]_4$ (*361*). The delocalized nature of the unpaired electrons in oxidized species has been confirmed by magnetic-field Mössbauer spectroscopy (*176, 177*) and by the very low activation energy for oxidations with $[CpFe(CO)]_4^+$ (*48*). Triangular clusters which have been oxidized and reduced are $Cp_3Fe_3(CO)_2S(SR)$ (*196*), $Co_3(CO)_9CX$ (*287*), and $Cp_3Co_3S_2$ (*178*). Only in one case was a one-electron oxidation accompanied by a chemical reaction. This occurred with the triangular $[Me_6C_6Nb_3Cl_6]^+$ species which can be oxidized chemically or electrochemically to the 2+ compounds (*243, 244*). These seem to be dimeric and contain octahedral Nb_6 clusters in which the two Nb_3 triangles are connected by one bonding electron pair. The facile redox chemistry of clusters is another demonstration of their catalytic qualities.

The third demonstration of this kind is the ease of hydrogenation of several carbonyl clusters, a reaction which may either be called a reduction to a metal hydride or an oxidative addition of hydrogen to the metal cluster. $Ru_3(CO)_{12}$ and $Os_3(CO)_{12}$ are hydrogenated by hydrogen in inert solvents either at high pressure and lower temperatures (*318, 398*), or even at atmospheric pressure and higher temperatures (*234, 251*). These discoveries have made the cluster hydrides $H_2Ru_4(CO)_{13}$, $H_4Ru_4(CO)_{12}$, $H_2Os_3(CO)_{10}$, and $H_4Os_4(CO)_{12}$ available in research quantities. The same holds true for the preparation of $H_3Re_3(CO)_{12}$ and $H_4Re_4(CO)_{12}$ from $Re_2(CO)_{10}$ (*234*). More interesting, substituted clusters are hydrogenated too. Thus, $Ru_3(CO)_{10}NPh$ is converted to $H_2Ru_3(CO)_9NPh$ (*337*), a compound of the type $H_2M_3(CO)_9Y$ mentioned in reaction (1). And olefin-substituted ruthenium and osmium clusters are hydrogenated under very mild conditions at the metals and at the olefin ligands (*74, 75, 131*). For instance, $HRu_3(CO)_9C_8H_{11}$ is converted to $H_2Ru_3(CO)_9C_8H_{12}$ (*74*), and $H_2Os_3(CO)_9$ ($C{=}CH_2$) goes to $H_3Os_3(CO)_9(CCH_3)$ (*131*). It has been proposed that H_2 addition to the metal atoms is the first step in these reactions (*74*). Their relation to catalytic processes is obvious.

The high prices of noble metals or catalyst losses have so far prevented the large-scale use of cluster catalysts. It is known, however, that $Rh_4(CO)_{12}$ and phosphine derivatives thereof are very active in hydroformylation (*88*), and that osmium clusters work efficiently in the cyclotrimerization of acetylenes (*371, 379*), as does $Ni_4(CNR)_7$ for acetylene or butadiene (*122*). Other processes found to be catalyzed by clusters are hydrogenations, isomerizations, and carbonylations by $Ru_3(CO)_{12}$ (*318*) and $[Co(CO)_2PR_3]_3$ (*322*), oxidations of CO-containing compounds such as CO and ketones with oxygen by $Rh_6(CO)_{16}$ (*291*), and radical polymerizations and dimerizations of olefins by methinyltricobalt enneacarbonyls (*235, 312*). The recent development of polymer-bound catalysts to combine the advantages of homogeneous and heterogeneous catalysis has also been applied to clusters, especially those of rhodium on polymeric materials with phosphine ligands (*104, 221*). This work provides the link between clusters and plain metals because catalytic activity is still shown by surface-bound materials resulting from $Rh_4(CO)_{12}$ and $Rh_6(CO)_{16}$, which are free from CO but can be recovered as the cluster carbonyls by subsequent addition of CO (*104, 364*). These surface-bound materials must be metal particles similar to those obtained from metal vapors on surfaces or by controlled decomposition reactions of simple metal compounds. The reactivity of these metal particles has not yet been investigated in detail, but as the mild addition of hydrogen to clusters has shown, it is not necessary to strip many ligands off in order to have a cluster show metal-like reactions.

3. Cluster Chemistry

In this chapter the chemistry of the individual clusters as reported from 1971 to 1975 will be dealt with according to the position of the metals in the periodic table. This part of the review is decidedly noncritical. According to the method chosen for the presentation of material certain repetitions of facts already mentioned in Chapter 2. cannot be avoided.

3.1. Groups V to VII

Compared to the inorganic cluster chemistry of the group V and VI metals, their organometallic cluster chemistry is poor. Of niobium and tantalum only one type of clusters was observed, namely $[Me_6C_6M_3Cl_6]^+$ (M = Nb, Ta) (*243, 244*), the structure of which (*93*) bears resemblance to the triangular chloro complexes of niobium. An interesting feature of these complexes is their oxidation which seems to lead to dimers in which the two triangular M_3 units are held together by just one electron pair.

Likewise, only one type of molybdenum cluster with organic ligands was reported. The complex [*29*] could be obtained as a cation from $Cp(CO)_3MoCl$ and $(Me_3Sn)_2S$ (*384*) or neutral from $HMo(CO)_2[P(OPh)_3]Cp$ and propylene sulfide (*35*).

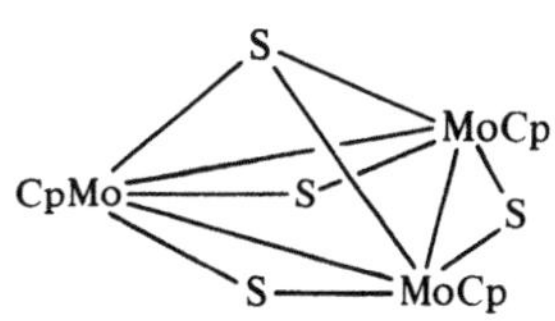

29

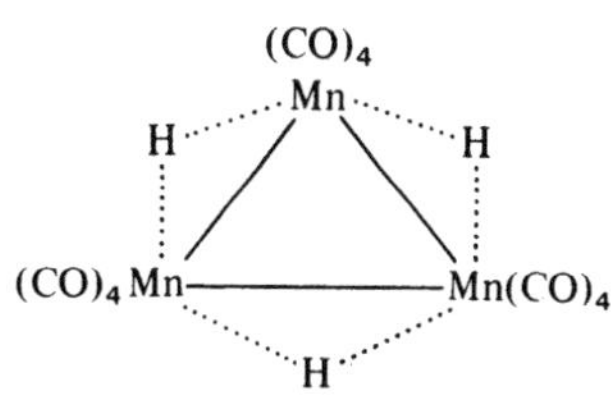

30

Of manganese, the structure of $Cp_3Mn_3(NO)_4$ (*158*) has been confirmed, and the structure of $H_3Mn_3(CO)_{12}$ [*30*], which is similar to that of $Os_3(CO)_{12}$, has been elucidated by inelastic neutron scattering (*394*) and X-ray crystallography (*246*). A new way of preparing the corresponding anion $H_2Mn_3(CO)_{12}^-$ from $Mn_2(CO)_{10}$ has been found (*161*).

A series of main-group transition metal clusters of manganese and rhenium with the structural types [*31*] and [*32*] has been isolated from reactions of the carbonyls with gallium and indium (*201, 202, 323, 324*).

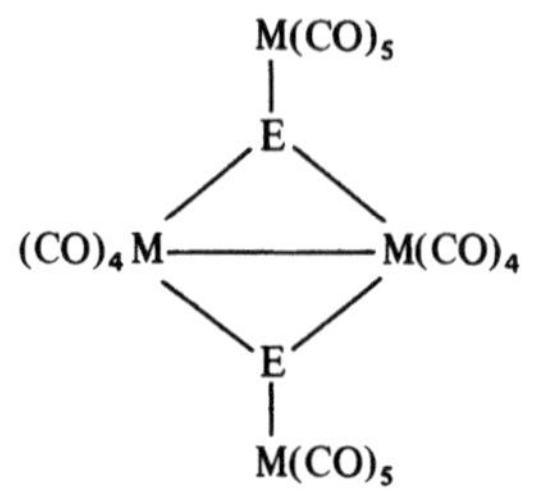

31

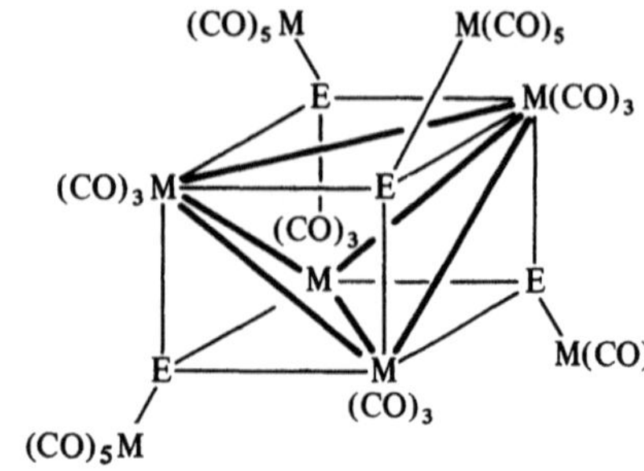

32

M = Mn,Re
E = Ga,In

The chemistry of pure rhenium clusters (*273, 332*) is confined in this context mainly to the hydrides. Two new preparations of $H_3Re_3(CO)_{12}$ were described (*204, 234*), the structure of which is like that of [*30*] (*394*); and $H_4Re_4(CO)_{15}^-$ of structure [7] was obtained from $Re_2(CO)_{10}$ and KOMe (*17*).

The phosphine-substituted cluster $Re_4(CO)_{10}(PMePh_2)_6$ resulted from $Re_2(CO)_{10}$ and the ligand under u.v. irradiation (*296*). A rare example of a sulfur atom bridging four metal atoms is present in the heteronuclear cluster [*33*] prepared from $[(CO)_4Re{-}S{-}SnMe_3]_2$ and $Cp(CO)_3MoCl$ (*383*).

33

3.2. *Iron*

The majority of iron clusters are triangular Fe_3 systems. Of these, the structure of the basic compound, $Fe_3(CO)_{12}$, was only recently established with certainty (*113*), and its fluxional properties, although investigated in some detail (*9, 111, 114, 175, 186*), are still not fully understood.

Clusters derived from $Fe_3(CO)_{12}$ by substitution of CO groups are few (*153*), because most nucleophiles cause degradation of the carbonyl. Only with isonitriles could facile substitutions be achieved (*192*), and the complex $Fe_3(CO)_9(PMe_2Ph)_3$ could be obtained under carefully controlled conditions and confirmed by a crystal structure (*325*). Chelating ligands of the ffars type replace two CO groups from $Fe_3(CO)_{12}$ initially (*84*), but react further with opening of Fe–Fe bonds to produce complexes like [*34*] (*156, 157*), whereas the related chelating ligand of [*35*] (*116*) is introduced into the $Fe_3(CO)_{10}$ unit via reaction with $Fe_2(CO)_9$.

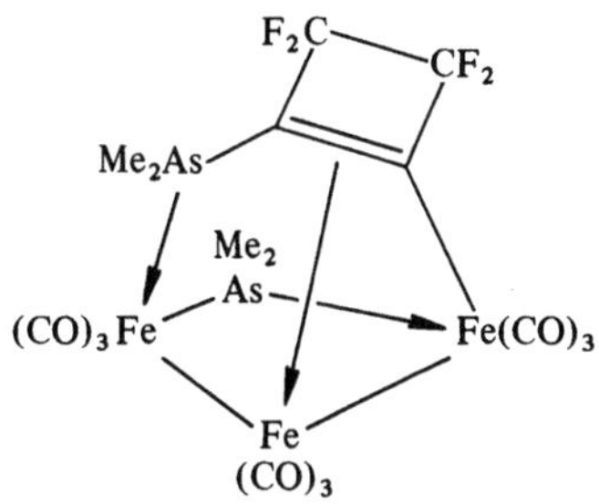

34

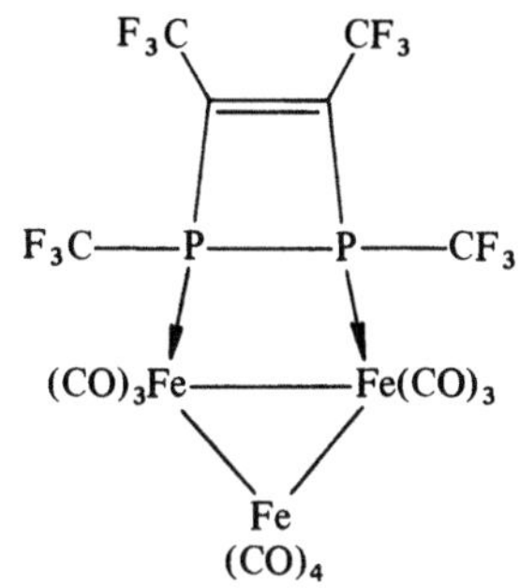

35

Another example of partial degradation of the $Fe_3(CO)_{12}$ triangle is furnished by its reactions with phosphino-acetylenes which lead to complicated open Fe_3 structures bridged by phosphido ligands and resulting acetylene dimers (*79, 311*).

Quite a number of new Fe_3 clusters has resulted from unusual reactions. Among these are the nitrogen bridged triangles [*36*] prepared from $Fe_2(CO)_9$ and silyl azides (*31, 255*) and [*37*] prepared from $Fe_2(CO)_9$ and several azo compounds (*24, 25*), while the previously known clusters $Fe_3(CO)_9(NR)_2$ were prepared from $Fe(CO)_5$ and organic nitro compounds (*26*).

Phenyl phosphine and $Fe(CO)_5$ at high temperatures form the open triangle of [*38*] (*374*), whereas the photolysis of $Cp(CO)_2Fe–PO(CF_3)_2$ yields a phosphorus-bridged cluster $Cp_2Fe_3(CO)_2PO(CF_3)_2$ of unknown structure (*141, 142*). Yet another cluster type [*39*] is formed from $Fe(CO)_5$ and AsF_3 (*135*).

Among the sulfur-bridged trinuclear iron carbonyls $S_2Fe_3(CO)_9$ (structure like [*38*]) has been prepared in a new way (*82*), and CO substitutions of this (*83*) and the

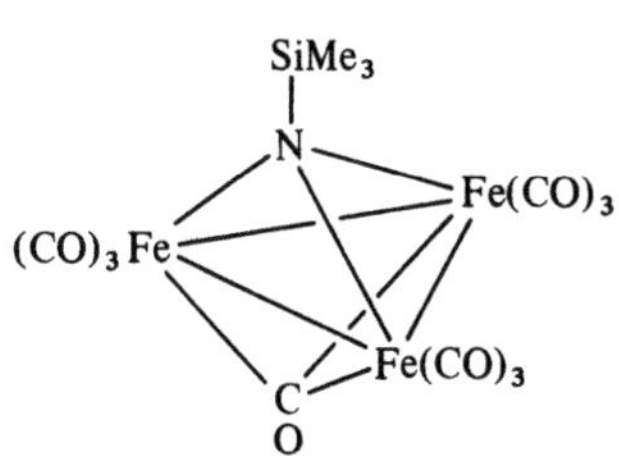

36

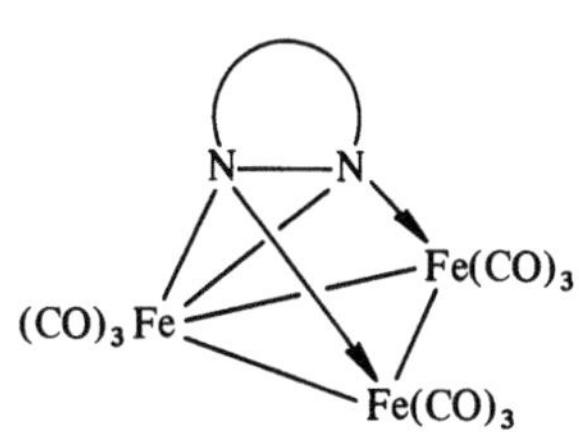

37

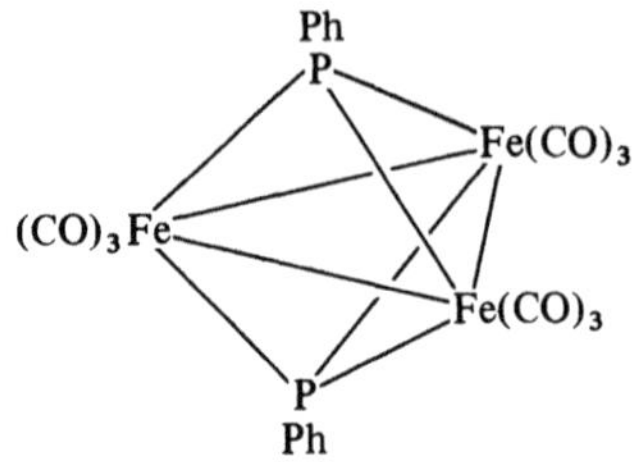

38

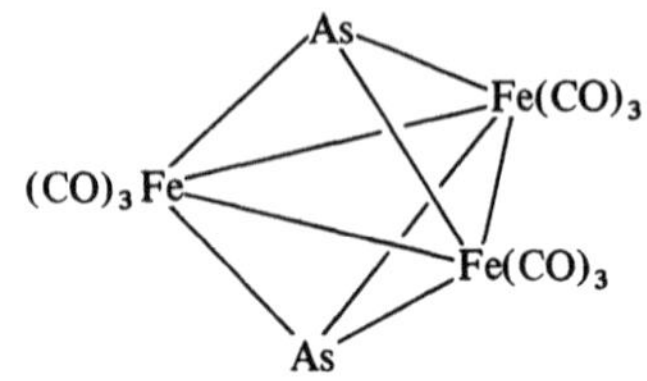

39

corresponding selenium and tellurium clusters (*331*) have been achieved. Unusual sulfur-bridged Fe_3 complexes were obtained from iron carbonyls and thioureas and thiobenzamides (*27*). From the reactions of $[CpFe(CO)_2]_2$ with disulfides the clusters $Cp_3Fe_3(CO)_2S(SR)$ were isolated (*196, 197*) which have the structure [*40*] (*381*). Triply-bridging sulfur is also present in the structure [*41*] (*34*) of $HFe_3(CO)_9SR$ ($R{=}i{-}C_3H_7$) obtained from $Fe_3(CO)_{12}$ and HSR. A different bonding situation with unusual doubly-bridging thioether ligands is present in $Fe_3(CO)_8(C_4H_8S)_2$ [*42*] prepared from $Fe_2(CO)_9$ and tetrahydrothiophene (*114*).

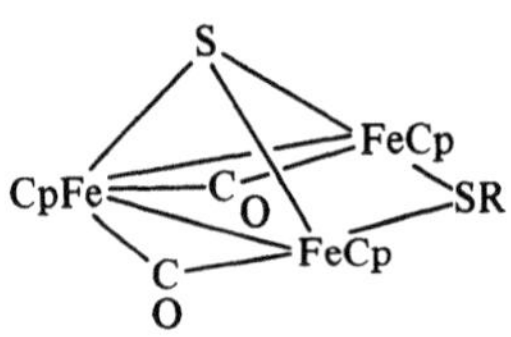

40

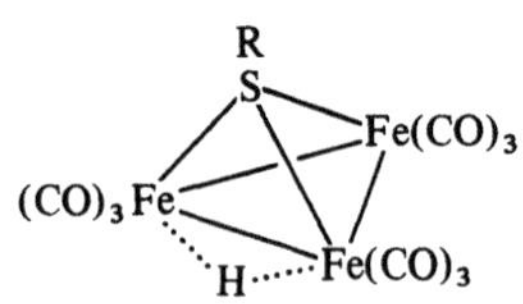

41

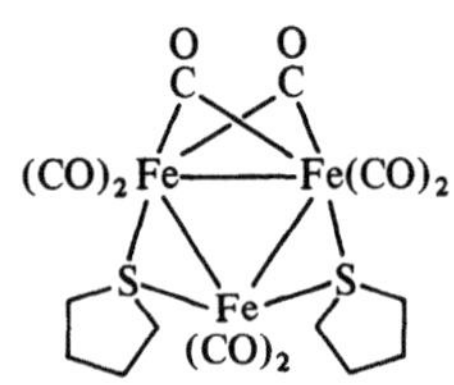

42

New types of carbon bridging occur in complexes [*43*] and [*44*]. [*43*] was obtained from $HFe_3(CO)_{11}^-$ and $MeOSO_2F$ as a source of cationic methyl groups (*360*). [*44*] (*402*) resulted either from $Cp(CO)_2Fe{-}C{\equiv}CPh$ (*403*) or its CuCl adduct (*6*) and $Fe_2(CO)_9$.

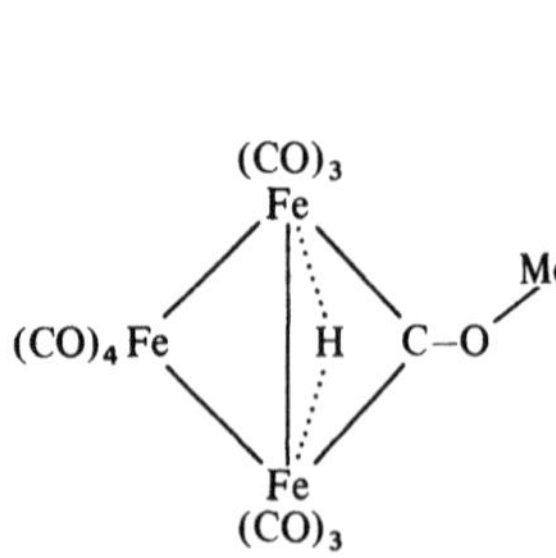

43

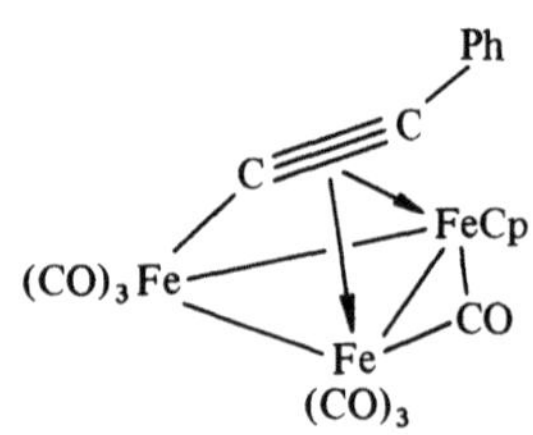

44

All the tetrameric iron clusters belong to the cubane class discussed in Chapter 2.2. Of these, $[CpFe(CO)]_4$ (*303*), $[(NO)FeS]_4$ (*180*) and $(NO)_4Fe_4S_2(NCMe_3)_2$ (*180*) have the totally bonding Fe_4 core, whereas $Fe_4S_4[S_2C_2(CF_3)_2]_4^{2-}$ (*41, 176*) belongs to the partially bonded group. The biological interest in $Fe_4(\mu-X)_4$ cubanes (*207*) has led to several spectroscopic investigations of these clusters (*177, 370*) and to studies of their redox behavior (*48, 166, 167, 375*). The triply-bridging CO groups in $[CpFe(CO)]_4$ are basic enough to bond to the Lewis acidic boron and aluminium halides (*266*).

Iron clusters of higher nuclearity are found only among the carbide and heterometallic types. $Fe_5C(CO)_{15}$ is readily accessible now (*367*), and up to three CO groups in this cluster can be replaced by phosphine ligands (*108*). $Fe_5C(CO)_{14}^{2-}$ of structure [*9*] was obtained in an attempt to prepare heteronuclear clusters from $Fe(CO)_5$ and $MoCp(CO)_3^-$ (*214*). $Fe_6C(CO)_{16}^{2-}$, a regular octahedral 86-electron cluster (*101*), was also obtained from a heterometallic reaction, this time between $Fe(CO)_5$ and $Mn(CO)_5^-$ (*102*). Another product from this reaction is triangular $MnFe_2(CO)_{12}^-$ of structure [*45*], of which phosphine-substituted derivatives could be prepared (*106*). And the tetranuclear heterometallic cluster [*46*] resulted from the reaction of $MCp(CO)_3^-$ (M = Mo, W) with $Fe_2(CO)_9$ (*215*).

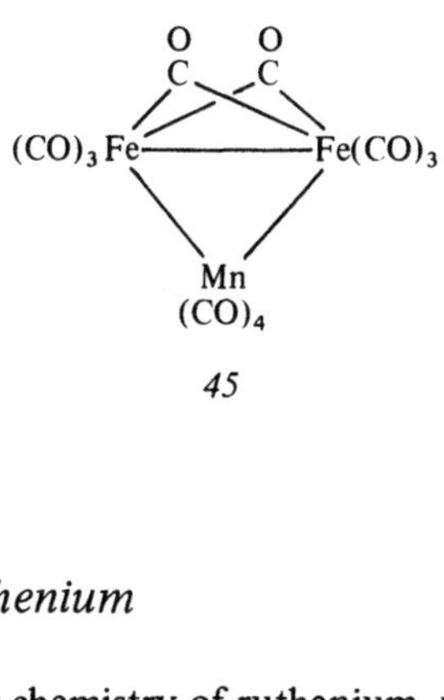

45

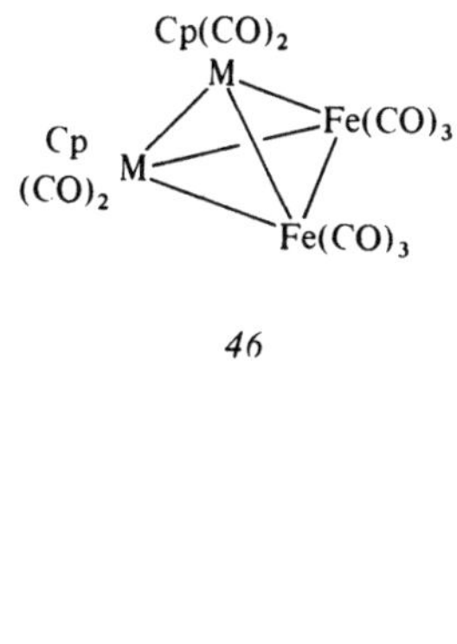

46

3.3. *Ruthenium*

The cluster chemistry of ruthenium, which is essentially carbonyl chemistry, has produced the highest number of publications including one review article (*376*). As is the case with iron, most ruthenium clusters are of the triangular Ru_3 type, but a considerable number of Ru_4 and Ru_6 clusters have also been investigated. The higher stability of ruthenium and osmium carbonyl clusters against degradation has enabled preparation of many derivatives with hydride, olefin, and Lewis basic ligands which could not be obtained from iron carbonyls because of the reaction conditions required.

The parent carbonyl $Ru_3(CO)_{12}$ can be obtained in small quantities from the decomposition of $H_2Ru(CO)_4$ (*398*) or by reduction of oligomeric (dien)$RuCl_2$ with zink in an atmosphere of CO (*245*), but in better yields and on a larger scale by reductions under CO of $RuCl_3 \cdot xH_2O$ (*121*) or of $Ru(acac)_3$ (*227*). Its fluxional properties have been investigated together with those of $Fe_2Ru(CO)_{12}$ (*9, 175*), and it has been found to form an adduct with $AlBr_3$ (*266*).

Several Lewis-base-substitution products of $Ru_3(CO)_{12}$ have been prepared (*153*). PF_3 replaces up to six CO groups (*377*), whereas organic phosphines or arsines replace two or three (*58, 65, 66, 230, 327*). The mechanism of these substitution reactions has been elucidated (*319–321*) and can be summarized by equation (2):

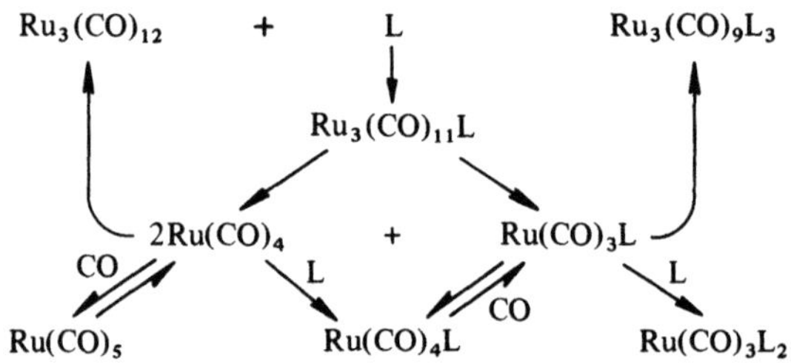

(2)

The substitution reactions can be accompanied by subsequent reactions. Thus, $Ru_3(CO)_{12}$ reacts with azobenzene (*61*) or fluorinated azobenzenes (*60*) to yield products like [*47*], and the pyrolysis of $Ru_3(CO)_9L_3$ complexes leads to reactions similar to those discussed in Chapter 3.4. for the corresponding osmium clusters. Rearrangements and orthometalations were observed (*65, 66*), and one cluster formulated as [*48*] was isolated (*65*).

Other triangular Ru_3 derivatives include $Ru_3(CO)_{10}(NO)_2$ (*308*), which has structure [*49*] with only two Ru–Ru bonds and allows substitution of two CO groups by phosphines (*307*). The germanium-bridged cluster [*50*] can be prepared from $Ru_3(CO)_{12}$ and Me_3GeH (*211*) or from $(Me_3Ge)_2Ru(CO)_4$ (*253*). Photolysis of $Cp(CO)_2Ru–SR$ yields $Cp_3Ru_3(CO)_2(SR)_3$ for which structure [*51*] was suggested

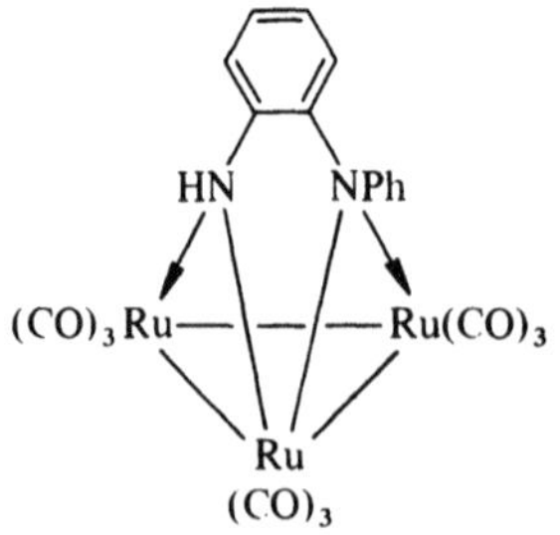

47

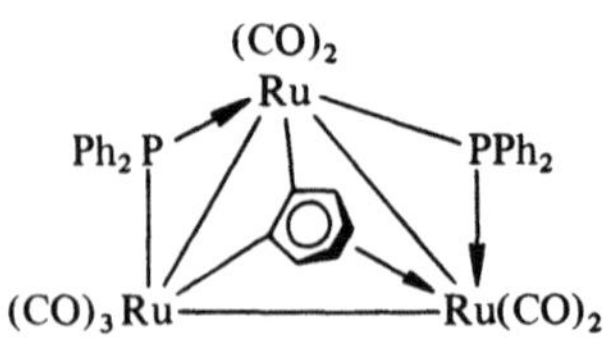

48

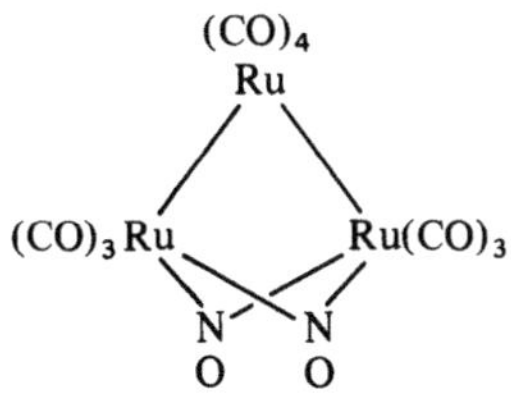

49

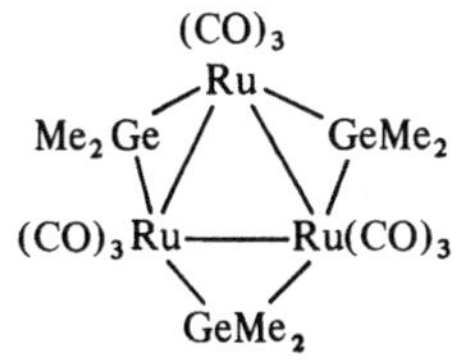

50

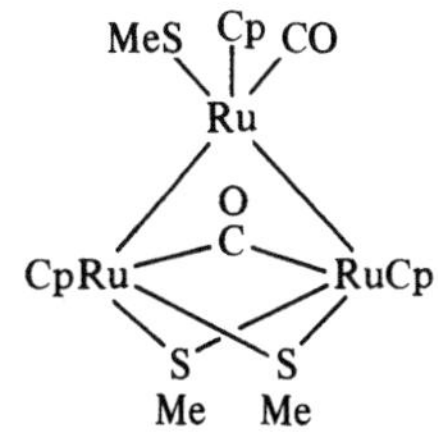

51

(*252*). And the reaction between $Me_3Si{-}N_3$ and $Ru_3(CO)_{12}$ leads to $Ru_3(CO)_9$-($\mu_3{-}CO$)($\mu_3{-}NSiMe_3$), which probably has a structure analogous to that of *36* (*1*).

Three types of triruthenium hydride clusters are known: $HRu_3(CO)_{10}X$ [*52*], $H_2Ru_3(CO)_9Y$ [*53*], and $H_3Ru_3(CO)_9Z$ [*54*]. Of type [*52*], the complex with X=C–NMe_2 was prepared from $Ru_3(CO)_{12}$ and $Me_3Sn{-}CH_2NMe_2$ (*95*) with X = SEt from $Ru_3(CO)_{12}$ and RSH (*123*), and with X = NHPh from $Ru_3(CO)_{12}$ and aniline or nitrobenzene (*337*). Of type [*53*], the complexes with Y = S, Se or Te were prepared from $Ru_3(CO)_{12}$ and YO_3^{2-} (*335*) or from $HRu_3(CO)_{10}SEt$ and concentrated H_2SO_4 (*123*), and the complex with Y = NPh from $Ru_3(CO)_{10}NPh$ and hydrogen under pressure (*337*). The complex [*54*] with Z = S^+ is an intermediate in the preparation of $H_2Ru_3(CO)_9S$ (*123*) and the complex [*54*] with Z = CMe was obtained from $H_4Ru_4(CO)_{12}$ and ethylene (*67, 76, 359*).

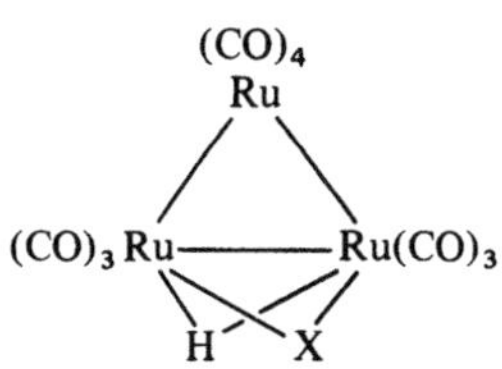

52

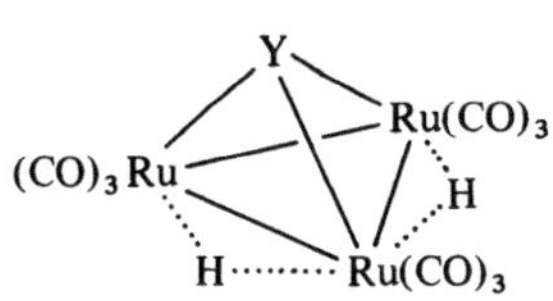

53

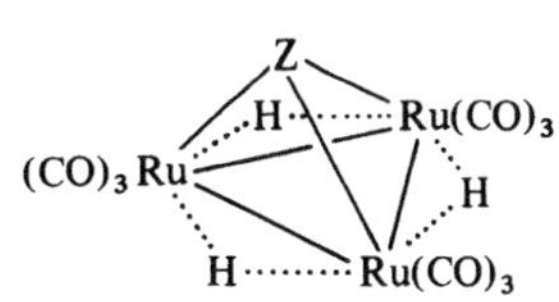

54

Numerous investigations have been undertaken on the reactions of ruthenium carbonyls with olefins and acetylenes. Two complex types [*55*] and [*56*] result from the reaction of $Ru_3(CO)_{12}$ with ethylene and other simple olefins (*127*). The complexes [*56*], which belong to the products of reaction (1) (Chapter 2.5.) are also formed from $Ru_3(CO)_{12}$ and diphenyl acetylene (*183*). [*55*] and [*56*] show interesting fluxional properties, and four different types of ligand scrambling are possible (*163*).

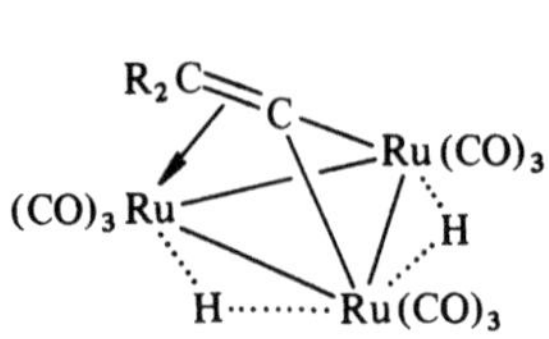

55

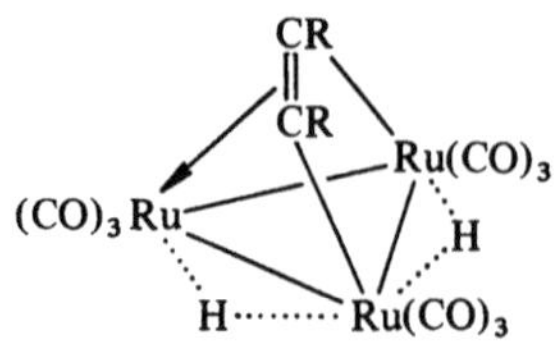

56

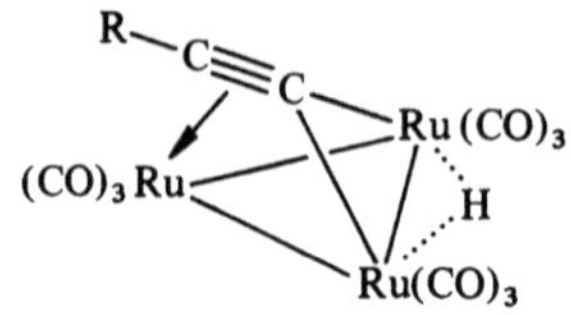

57

Terminal acetylenes and $Ru_3(CO)_{12}$ yield complexes of the type [*57*] (*9*, *190*, *336*), whereas internal acetylenes form either complexes [*56*] or acetylene-substituted Ru_4 complexes (*229*). Alternatively, two acetylene moieties are incorporated with formation of metallacyclopentadienes (*229*), a class of compounds more familiar in osmium cluster chemistry (cf. Chapter 3.4.). Instead of two acetylene molecules, one molecule of an arylbutadiene may be the precursor of the metallacycle (*382*).

Linear and cyclic diolefins form a variety of clusters with trirutheniumdodecacarbonyl or the tetraruthenium carbonyl hydrides. These belong to the compound types $H_2Ru_3(CO)_9L$ [*56*] or $HRu_3(CO)_9L'$ [*58*]. Type [*56*] complexes are obtained from cyclopentadienes (*217*), cyclooctadienes (*74*), or bicyclooctadiene (*146*). Type [*58*] complexes are obtained from cyclooctadienes (*74*), too, but also from cyclododecatrienes (*59*, *117*), hexadienes (*165*), and butadienes (*185*). Some of these reactions also lead to tetraruthenium olefin clusters. Of these, the complex [*1*] has been mentioned already, and is obtained from cyclooctadiene (*74*, *75*, *283*), whereas the complex [*59*] results from cyclododecatriene (*36*, *59*).

Cycloheptatrienes and $Ru_3(CO)_{12}$ yield yet another class of Ru_3 clusters which may be formulated as [*60*] (*32*, *68*), and cyclooctatetraene forms the similar complex [*61*] (*33*).

In certain cases cyclization of cyclooctatetraene occurs with $Ru_3(CO)_{12}$, and pentalene complexes [*62*] are formed (*209*, *210*), which are similar but not structurally related to the complexes [*63*] resulting from azulene and $Ru_3(CO)_{12}$ (*96*, *100*).

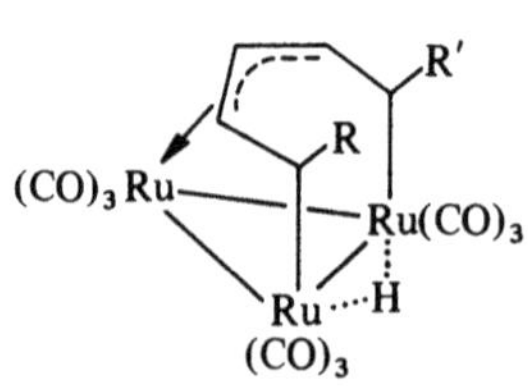

58

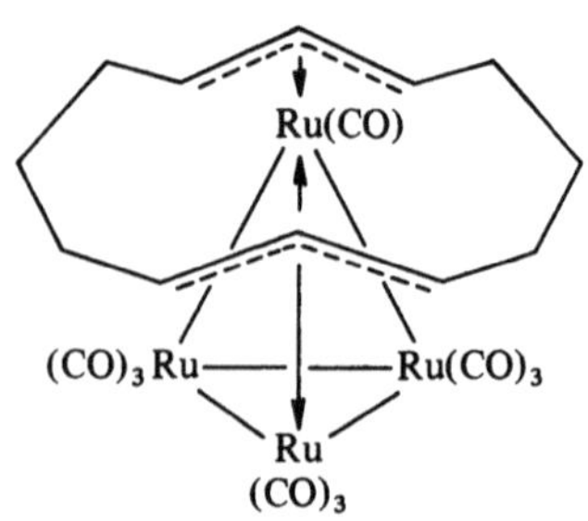

59

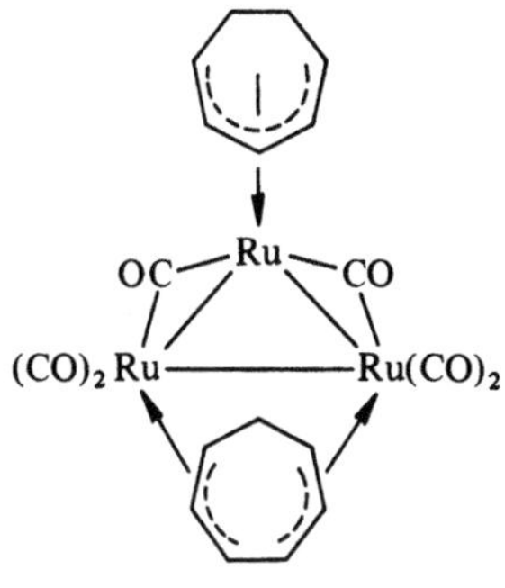

60

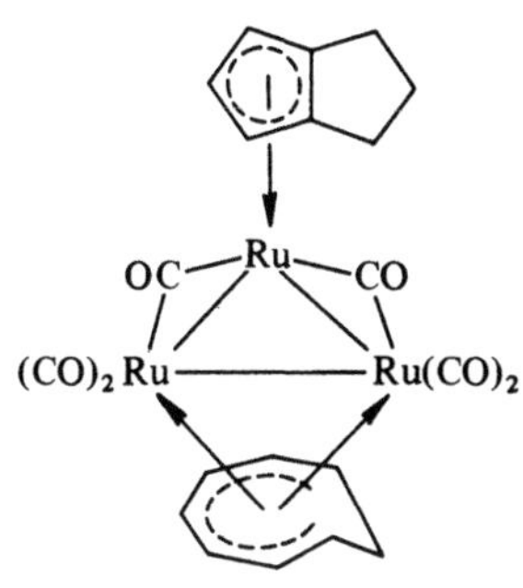

61

The chemistry of the tetranuclear carbonyl hydrides of ruthenium has been developed since they have become available by novel syntheses. $H_4Ru_4(CO)_{12}$, $H_2Ru_4(CO)_{13}$, and substituted derivatives thereof are formed as by-products in such reactions as the decomposition of $H_2Ru(CO)_4$ (*318, 398*), the substitution of $Ru_3(CO)_{12}$ with aryl phosphites (*66*) the reaction of $Ru_3(CO)_{12}$ with aryl butadienes (*382*) or manganese and rhenium carbonyl anions (*248, 249*), the pyrolysis of $Ru_3(CO)_{12}$ in the presence of water (*150*), and hydrogenations of olefin triruthenium clusters (*74*) or $Ru_3(CO)_{10}NPh$ (*337*). Hydrogenation at high temperatures and ordinary pressure is also the method of choice for the larger-scale synthesis of $H_4Ru_4(CO)_{12}$ (*234, 251*). The structure of tetrahedral $H_2Ru_4(CO)_{13}$ has been determined (*404*), and hydrogen-atom fluctuation has been observed in $H_4Ru_4(CO)_{12-n}$-$[P(OMe)_3]_n$ (*250*) and $H_3Ru_4(CO)_{12}^-$ (*254*). A considerable number of the ruthenium olefin clusters mentioned above were prepared starting from the hydride clusters.

Few clusters with more than four ruthenium atoms are known. Among these is $H_2Ru_6(CO)_{18}$, an 86 electron octahedral cluster with face-bridging hydrogen ligands (*99*), which can be prepared from $Ru_3(CO)_{12}$ and $Mn(CO)_5^-$ or $Re(CO)_5^-$ followed by acidification (*248, 249*). Pyrolysis of $Ru_3(CO)_{12}$ in the absence of solvent leads to $Ru_5C(CO)_{15}$ and $Ru_6C(CO)_{17}$ (*149, 151*) which are also obtained from

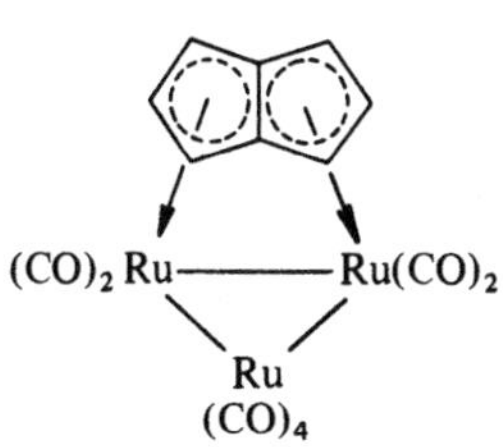

62

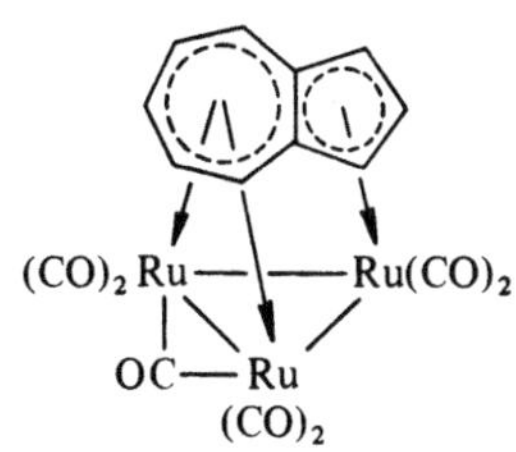

63

$H_4Ru_4(CO)_{12}$ (*152*). For $Ru_5C(CO)_{15}$, the tetragonal pyramidal Ru_5 unit, which was proved for $Fe_5C(CO)_{15}$, was deduced from the IR spectra. $Ru_6C(CO)_{17}$, which is isoelectronic with $Fe_6C(CO)_{16}^{2-}$, contains the carbon atom in the center of the Ru_6 octahedron.

Of the heterometallic ruthenium clusters, some are among the many reaction products of $Ru_3(CO)_{12}$ and $Mn(CO)_5^-$ or $Re(CO)_5^-$. These are $ReRu_3(CO)_{16}^-$, which was assigned a structure [*5*] or [*8*] like that of the osmium analogs (Chapter 2.2.), $H_2Re_2Ru(CO)_{12}$, and $H_2Re_2Ru_2(CO)_{16}$, for which no structures were suggested (*248, 249*). The reaction between $Fe_2(CO)_9$ and $Cp(PPh_3)_2Ru{-}C{\equiv}CPh$ or its CuCl adduct yields [*64*] (*7, 57*) which is similar to [*44*]. And of another Fe–Ru cluster, $H_2FeRu_3(CO)_{13}$ the fluxional properties (*293*) and the structure [*65*] (*191*) were determined.

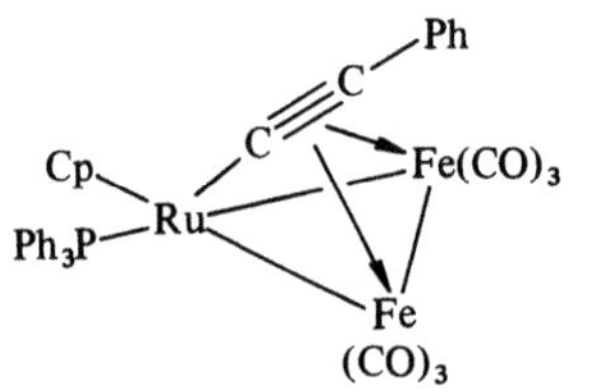

64

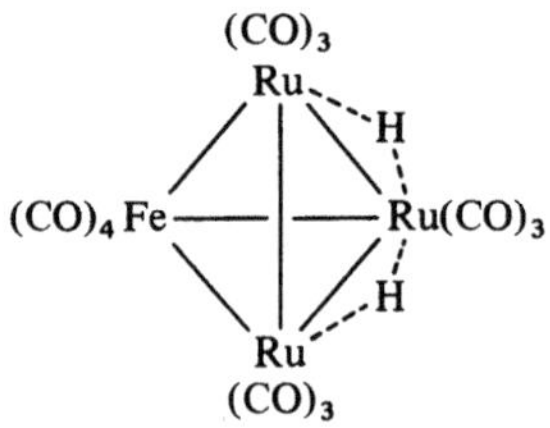

65

3.4. Osmium

As in the case of ruthenium, osmium cluster chemistry is carbonyl chemistry (*376*). But the greater stability of the Os_3 unit against nucleophilic degradation has enabled the preparation of a larger number of variously substituted trisomium carbonyls and to investigate their reactions under much more forceful conditions.

New good yield syntheses of $Os_3(CO)_{12}$ by carbonylation of OsO_4 (*227*) and of $H_2Os_3(CO)_{10}$ and $H_4Os_4(CO)_{12}$ by hydrogenation of $Os_3(CO)_{12}$ (*234, 251*) have been developed. The structure [66] of $H_2Os_3(CO)_{10}$ (*281*) is interesting as it contains an Os–Os double bond. The compound is converted by CO to $H_2Os_3(CO)_{11}$ (*124*) with structure [*67*] (*357*). Ligand scrambling in $Os_3(CO)_{12}$ is less facile than in $Fe_3(CO)_{12}$ and $Ru_3(CO)_{12}$ (*9, 175*), and the compound is basic enough to be protonated to $HOs_3(CO)_{12}^+$ with the hydrogen atom in an edge-bridging position (*394*). Other simple Os_3 derivatives are the silicon- and germanium-bridged clusters $[Me_2EOs(CO)_3]_3$ (E = Si, Ge), which are formed from silicon and germanium containing osmium carbonyls (*55, 253*), and which have structures like their ruthenium analog [*50*], and the nitrosyl-bridged $Os_3(CO)_{10}(NO)_2$, which is chemically and structurally analogous to its ruthenium counterpart [*49*] (*308*).

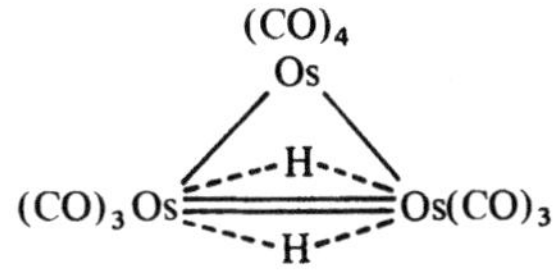

66

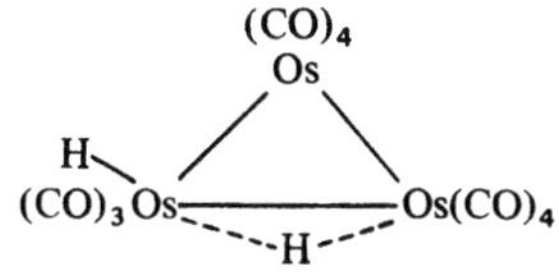

67

The substitution chemistry of triosmium carbonyls is extensive (*153*). Under mild conditions phosphines and arsines just replace CO ligands in $Os_3(CO)_{12}$ to form $Os_3(CO)_{12-n}L_n$, where L is PPh_3 (*50*), PMe_2Ph or $AsMe_2Ph$ (*128, 129*), PMe_3 or PEt_3 (*132*), and $C_4F_4(PPh_2)_2$ ($f_4$fos) (*118*). Similarly, $H_2Os_3(CO)_{10}$ is converted to $H_2Os_3(CO)_{10}L$ with structures similar to [*67*] by CO, PR_3, AsR_3, PhCN, and RNC (*124, 357*). Heating of these products, however, leads to ligand rearrangements, ligand fragmentations, and ortho-metalations, which are not generally predictable but which retain the Os_3 skeleton. Thus, the pyrolysis of the $Os_3(CO)_{12-n}(PPh_3)_n$ complexes leads to six new clusters [*68–73*] (*50, 51, 179*).

Another series of Os_3 clusters containing the triply-bridging ortho-phenylene group was obtained similarly from $Os_3(CO)_{12-n}(EMe_2Ph)_n$ (E = P, As) (*128, 129*). These complexes can be formulated as $XYOs_3(CO)_nC_6H_4$, one example being [*26*], and some of them are structurally related to [*68–73*]. A general pattern seems to emerge from these reactions which can be simplified by equation (1), Chapter 2.5.: In many cases the oxidative addition of free or complex bound substrates H_2Y leads to the clusters $H_2Os_3(CO)_9Y$ (*130, 132, 406*), and so widely differing ligands Y as in [*74–80*] are stabilized in the Os_3 cluster.

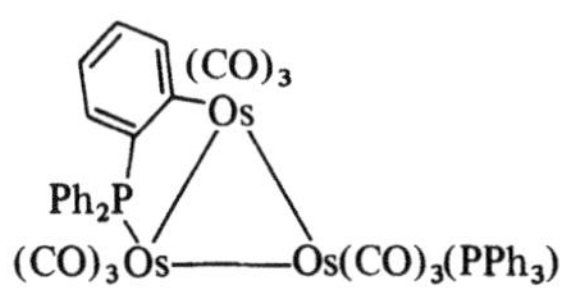

$HOs_3(CO)_9(PPh_3)(PPh_2C_6H_4)$

68

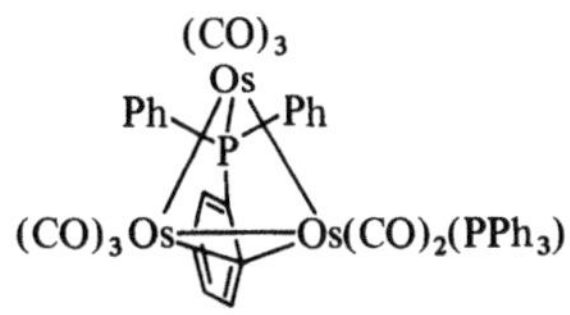

$HOs_3(CO)_8(PPh_3)(PPh_2C_6H_4)$

69

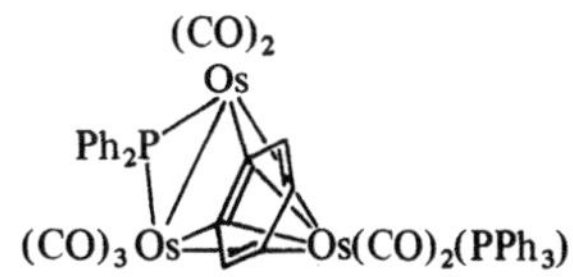

$HOs_3(CO)_7(PPh_2)(PPh_3)(C_6H_4)$

70

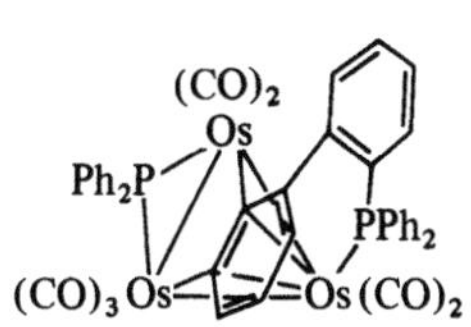

$HOs_3(CO)_7(PPh_2)(PPh_2C_6H_4C_6H_3)$

71

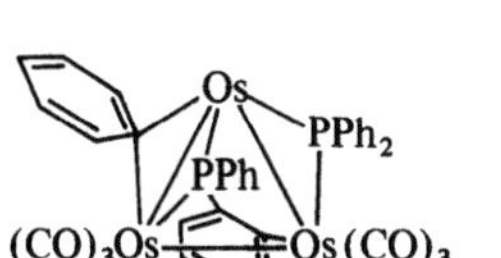

$Os_3(CO)_8(PPh_2)(Ph)(PPhC_6H_4)$

72

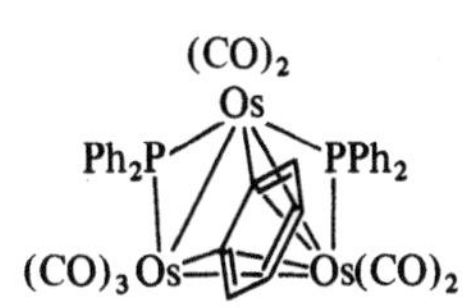

$Os_3(CO)_7(PPh_2)_2(C_6H_4)$

73

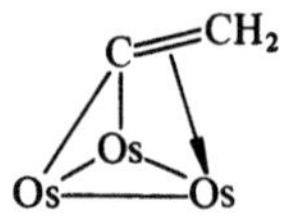

74

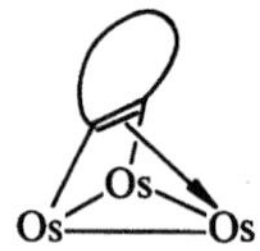

75

76

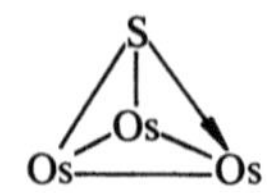

77

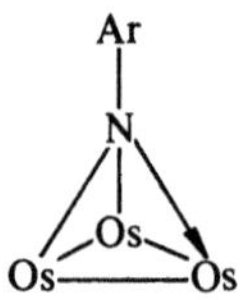

78

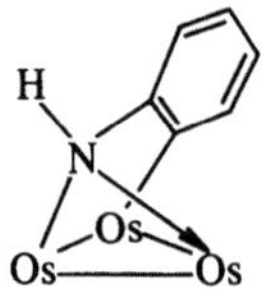

79

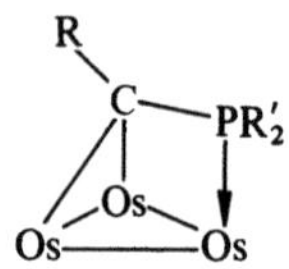

80

The first step of these ortho-metalation or oxidative-addition reactions which would lead to the complexes $HOs(YH)(CO)_{10}$ has been seen only in the case of H_2Y = anilines (*132, 406*) where [*81*] could be isolated. But similar complexes, [*82*] (*281*) and [*83*] (*407*), have also been obtained from $Os_3(CO)_{12}$ and EtSH or pyridine.

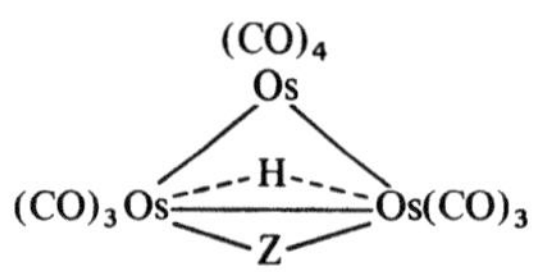

81 Z = YH = NHPh

82 Z = SEt

83 Z = o–NC_5H_4

Simple olefins like ethylene and substituted ethylenes react with $Os_3(CO)_{12}$ to the isomeric clusters $H_2Os_3(CO)_9(R_2C_2)$ with structures like [*55*] and [*56*] (*127, 133*). These reactions also follow equation (1) and product type [*55*] corresponds to [*74*] and [*56*] to [*75*]. Cyclopentene, cyclopentadiene, and benzene give complexes like [*56*] (*133, 217*). The fluxional properties of several osmium clusters of type [*55*] and [*56*] have been investigated (*163, 224*) and found to be like those of the corresponding ruthenium clusters. $H_2Os_3(CO)_9(CCH_2)$ can be protonated to [*84*] (*127*) and hydrogenated to $H_3Os_3(CO)_9(CCH_3)$ (*131*), the structure [*85*] of which (*405*) is analogous to that of the ruthenium complex [*54*].

The reactions to $H_2Os_3(CO)_9(C_2R_2)$ should also go via an intermediate, $HOs_3(CO)_{10}(CR{=}CHR)$ [*86*]. Such intermediates could not be isolated in reactions with $Os_3(CO)_{12}$, but the milder reaction conditions required by $H_2Os_3(CO)_{10}$ allowed it to react with olefins or acetylene, with the formation of [*86*] (*125, 126, 222, 239*). Heating of complexes [*86*] converts them to clusters like [*55*] and [*56*].

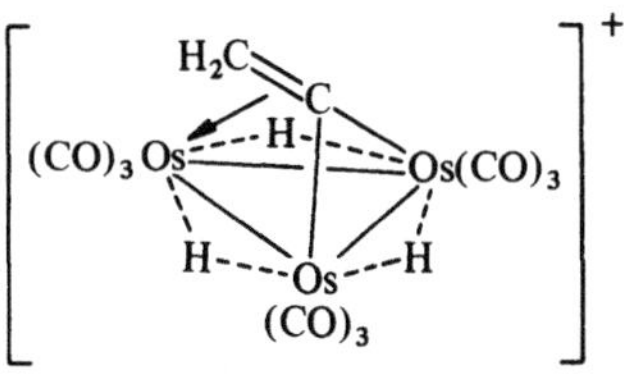

84

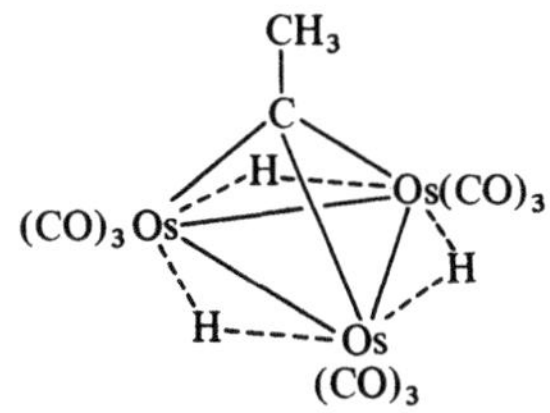

85

The interesting fluxional behavior of complexes [*86*] (*358*) includes oscillation of the olefinic ligand as in [*25a*] ⇄ [*25b*]. Reactions of $H_2Os_3(CO)_{10}$ with acetylenes also yield the complexes [*87*] (*126, 371*), which in the case of R=H can rearrange to the complex $HOs_3(CO)_9(C{=}CR)$ (*9, 125*) the ruthenium analog [*57*] of which is known. Thus, a series of reversible oxidative addition reactions of C–H bonds to osmium clusters has been established which should bear some relevance to catalytic processes.

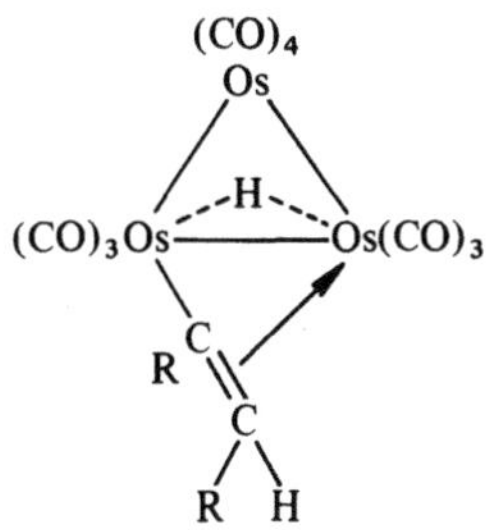

86

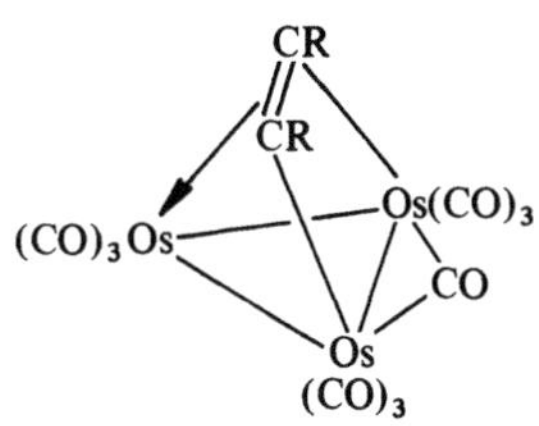

87

In many cases the reaction of osmium carbonyls and acetylenes does not stop at the first stages as in [*86*], [*87*], or [*57*]. Instead, two or more acetylene molecules are incorporated, and in some cases acetylene trimerization to benzenes takes place (*182, 371, 379*). Incorporation of two acetylene molecules can lead to metallacyclopentadiene clusters like [*88*] (*126, 168, 171, 182, 184, 223, 371*), or to metallacyclohexadienone clusters like [*89*] (*126, 223*). And the complex [*90*], another intermediate, is related to [*88*] by an intramolecular oxidative addition reaction (*168, 169*).

Incorporation of a third acetylene molecule takes place by CO replacement and without interference with the metallacyclopentadiene ring (*170, 371, 379*). In the cluster, then, the three acetylene ligands rearrange to a triacetylene ligand of unknown structure before the benzene is liberated (*371, 379*).

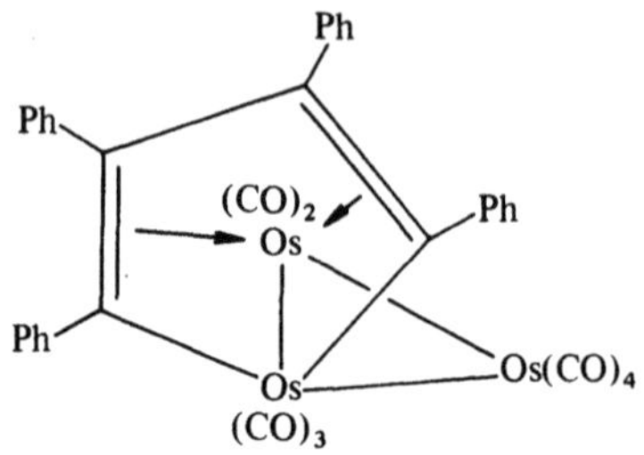

88

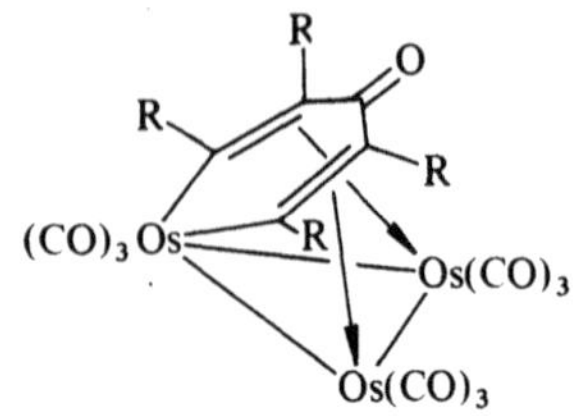

89

Larger osmium carbonyl clusters are obtained by pyrolysis reactions of $Os_3(CO)_{12}$ (*149, 151*). $Os_5(CO)_{16}$ is supposed to have a trigonal bipyramidal cluster framework; $Os_5C(CO)_{15}$ was shown to have the same structure as the corresponding iron and ruthenium clusters (*152*). $Os_6(CO)_{18}$, of which the fluxionality (*148*) and the crystal structure (*284*) were determined, has the unusual structure [*12*]. $Os_7(CO)_{21}$ has structure [*13*] (*151, 284*), whereas the structures of $Os_8(CO)_{23}$ and $Os_8C(CO)_{21}$ are still unknown. For all the known structures of these clusters, the skeletal electron-counting scheme (see Chapter 2.3.) predicted the correct metal atom arrangements (*151*). Pyrolysis of $Os_3(CO)_{12}$ in the presence of water yielded the series of osmium hydride clusters $HOs_3(CO)_{10}OH$, $H_2Os_4(CO)_{13}$, $H_4Os_4(CO)_{12}$, $H_2Os_5(CO)_{16}$, $H_2Os_5(CO)_{15}$, $H_2Os_6(CO)_{18}$, and $H_2Os_7C(CO)_{19}$ (*150*). And finally, the high-temperature carbonylation of OsO_4 besides $Os_3(CO)_{12}$ gave a small amount of $Os_6O_6(CO)_{16}$ of unknown structure (*49*).

Only one reaction type leading to heterometallic osmium clusters has been reported (*248, 249*). Among the products obtained from $Os_3(CO)_{12}$ and $M(CO)_5^-$ (M = Mn, Re) with subsequent acidification are $MOs_2(CO)_{12}^-$, for which a structure like that of $Os_3(CO)_{12}$ was suggested, and the corresponding hydride $HMOs_2(CO)_{12}$. For the triosmium clusters also obtained from this reaction the structures [*5*] and [*8*] were suggested for $HMOs_3(CO)_{15}$ and $HMOs_3(CO)_{16}$, while $H_3MOs_3(CO)_{13}$ is supposed to be structurally related to $Fe_4(CO)_{13}^{2-}$.

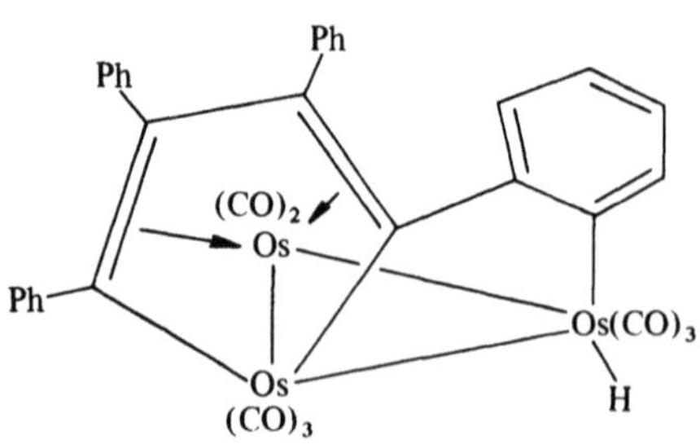

90

3.5. Cobalt

Cobalt offers many possibilities of cluster-core-geometry, but the chemistry of cobalt clusters is limited, again due to the weakness of first row metal-metal bonds and their susceptibility to nucleophilic cleavage. Only in case of the methinyl tricobalt ennea-carbonyls has a singular chemistry been developed, and therefore these compounds will be treated under a separate heading.

There are only two simple tricobalt clusters, namely $[CpCo(CO)]_3$, obtained by photolysis of $CpCo(CO)_2$ (*386, 387*), and $[(R_3P)Co(CO)_2]_3$ obtained from $[(R_3P)Co(CO)_3]_2$ under hydroformylation conditions (*322*). Both are found to contain three bridging CO groups.

All other trinuclear cobalt clusters contain triply-bridging ligands. Most of these belong to the structural type *91*, which is similar to the methinyl tricobalt enneacarbonyls. Structure *91* is realized with X = S or Se (*368*) (see [*27*] in Chapter 2.3.), PR (*276*), PS (*385*), and $B \cdot NEt_3$ (*339a*), of which the compounds with X = S, Se and PR, belong the the rare examples of paramagnetic metal carbonyls.

Only in the case of X = Al the structure contains bridging CO ligands (*341*). Trinuclear cyclopentadienyl cobalt clusters exist in the structural type [*92*] with two triply bridging ligands. They are formed with X = Y = NO (*301*) or X = Y = S for which case a cationic complex was prepared too (*178*), and for X = S and Y = CO (*178*) or X = $NSiMe_3$ and Y = CO (*1*).

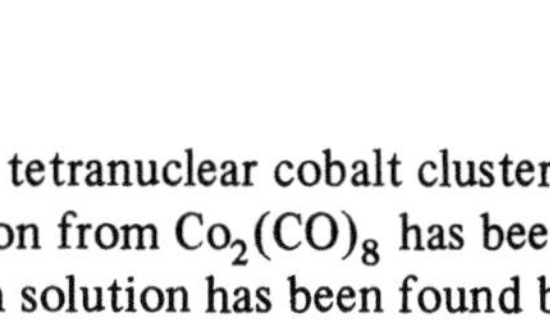

91

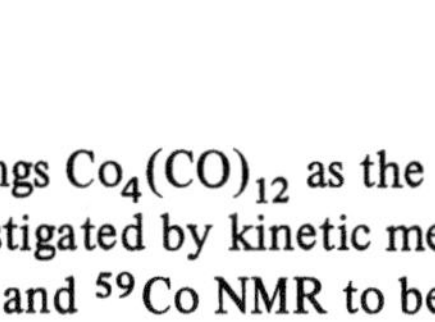

92

To the tetranuclear cobalt clusters belongs $Co_4(CO)_{12}$ as the simplest compound. Its formation from $Co_2(CO)_8$ has been investigated by kinetic methods (*378*) and its structure in solution has been found by ^{13}C and ^{59}Co NMR to be the same as in the solid state (*103, 162*). Substitution reactions proceed smoothly with up to four isocyanide ligands (*304*), and arene ligands replace three CO groups (*47, 240*) at the apical cobalt atom (*43, 162*). Only few phosphine substitution reactions of $Co_4(CO)_{12}$ have been reported (*268*), and it has been found that the first PR_3 ligand replaces a CO group at the basal cobalt triangle (*103*). Arene-$Co_3(CO)_9$ complexes also allow replacement of the arene or of one CO group by phosphines (*363*), whereas the cluster $Co_4(CO)_8(ffars)_2$, which contains the tetrahedral Co_4 skeleton (*154*), was not obtained from $Co_4(CO)_{12}$ but in an unexpected reaction from $CF_3CCo_3(CO)_9$ and ffars (*119*). Polyolefines normally destroy the $Co_4(CO)_{12}$ cluster, and only in the case of cycloheptatriene could mono- and disubstitution be achieved (*247*).

Other tetracobalt clusters are $[CpCoH]_4$, in which the hydrogen atoms bridge the faces of the cobalt tetrahedron (*219*), and which was obtained by borohydride reduction of $[CpCo(NO)]_2$ (*298*), and $Cp_4Co_4(CO)_2$ with two triply-bridging CO groups, again obtained from photolysis of $CpCo(CO)_2$ (*386, 387*). Among the cubane type Co_4X_4 clusters (see Chapter 2.2.) $[CpCoS]_4$ with structure [*21*] is a totally non-bonding case, but its cation shows partial metal-metal bonding (*361*). $[CpCoP]_4$ has structure [*22*] with two cobalt-cobalt bonds (*362*); and $[(NO)Co(NCMe_3)]_4$ is similar to structure [*23*] with four bonding and two nonbonding cobalt-cobalt interactions (*181*).

Few tetracobalt complexes are known which do not contain the tetrahedral cobalt atom arrangement. To these belong the acetylene clusters $Co_4(CO)_{10}(RC{\equiv}CR)$ with the butterfly structure like [1]. A new synthesis of this class of compounds starting from arene $Co_4(CO)_9$ and acetylenes has been found (*363*), and ^{13}C–NMR studies have revealed that above $-70°$ total CO scrambling occurs, while the acetylenic ligand remains rigid even at higher temperatures (*10, 162*). A planar rectangle of cobalt atoms was found in the structure *6* of the cluster $(PhP)_2Co_4(CO)_{10}$ prepared from $PhPCl_2$ and $NaCo(CO)_4$ (*334*), which is structurally related to $S_2Co_4(CO)_{10}$ and $Te_2Co_4(CO)_{10}$. The reaction between $Co_2(CO)_8$ and tetramethyldiphosphine disulfide leads to a cluster of composition $SCo_4(CO)_9(PMe_2)_2$ for which the spectra offer a choice of four possible structures (*302*).

Of the hexanuclear cobalt clusters, the highly symmetrical $Co_6(CO)_{14}^{4-}$ was the subject of a Wolfsberg-Helmholz MO calculation, which gave a theoretical background to the skeletal electron-counting scheme outlined in Chapter 2.3. (*294*). The reaction of $Co_3(CO)_9CCl$ with $NaCo(CO)_4$ yielded the carbido cluster anion $Co_6C(CO)_{15}^{2-}$, which is supposed to have the same structure as its rhodium analog with the carbon atom in the center of a trigonal prism of cobalt atoms (*14*).

Heterometallic cobalt clusters are known with several other metal atoms (see later chapters). Here the ones with iron should be mentioned. From mixtures of iron and cobalt carbonyls and chalcogen compounds the clusters $EFeCo_2(CO)_9$ with E = S, Se, Te were isolated. Their structures [*28*] (*366, 368*) are similar to those of the $ECo_3(CO)_9$ clusters [*27*] and thus offer the possibility for the experimental bonding investigation mentioned in Chapter 2.3. Of $SFeCo_2(CO)_9CO$ substitutions by isonitriles have been observed (*304*). Other trinuclear iron-cobalt clusters include $CpCoFe_2(CO)_9$ obtained from $CpCo(CO)_2$ and $Fe_2(CO)_9$, for which structure [*93*] was made plausible (*212*), and $CpFeCo_2(CO)_6(C{\equiv}CPh)$, for which structure [*94*] was suggested (*403*).

For $HFeCo_3(CO)_{12}$ inelastic neutron scattering indicated a structure with the hydrogen atom bound terminally to the iron atom (*394*), whereas a crystal-structure determination of $HFeCo_3(CO)_9[P(OMe)_3]_3$ revealed an apical $Fe(CO)_3$ group, an overall geometry like that in $Co_4(CO)_{12}$ and the location of the hydrogen atom bridging the Co_3 face (*216*). This compound, like similar ones with one, two, or three phosphine ligands, was obtained by phosphine substitution of $HFeCo_3(CO)_{12}$

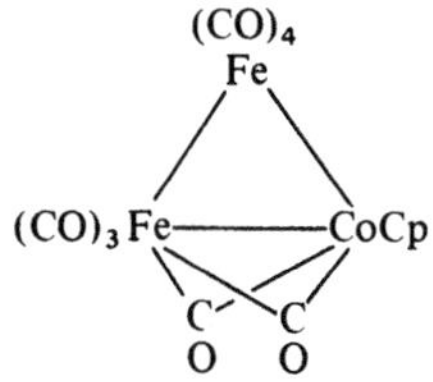

93

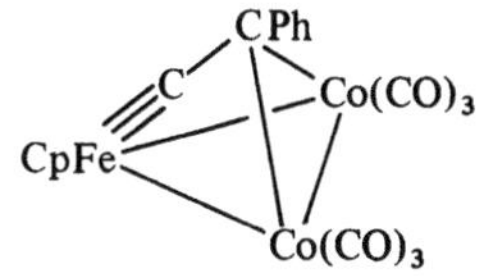

94

(*107, 216*). Similar substitutions could also be achieved with $FeCo_3(CO)_{12}^-$ where one or two CO groups could be replaced by phosphines (*106*). And diphenyl acetylene and $FeCo_3(CO)_{12}^-$ gave a complex $FeCo_3(CO)_{10}(Ph_2C_2)^-$ in which the metal atom tetrahedron is supposed to remain intact (*106*).

3.6. Methinyltricobalt Enneacarbonyls

The unique stability of the Co_3C cluster [95] compared to other triangular and tetrahedral clusters seems to result from the clamp effect of the triply-bridging methinyl ligand as well as from extensive electron delocalization. This stability has allowed numerous investigations of the organic chemistry and the coordination chemistry of the methinyltricobalt enneacarbonyls, which were summarized in two reviews (*313, 317*), the latest (*317*) of which covers the literature up to 1971. But since then enough material has been accumulated which extends our knowledge of the rich chemistry of these clusters.

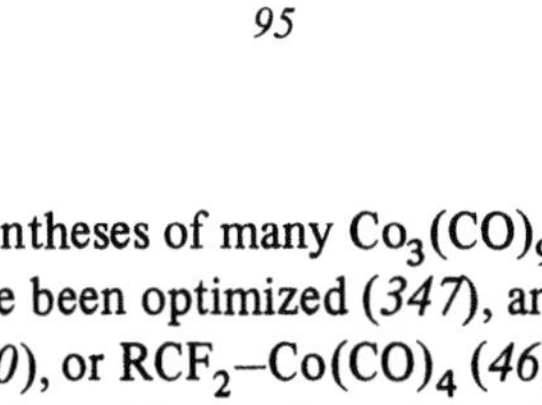

95

The syntheses of many $Co_3(CO)_9X$ compounds from dicobalt octacarbonyl and $XCCl_3$ have been optimized (*347*), and further reactions starting from $Co(CO)_4^-$ and $XCCl_3$ (*350*), or RCF_2–$Co(CO)_4$ (*46*) have been investigated in order to determine the mechanism of formation of the clusters. Methinyltricobalt enneacarbonyls are also formed from $Co_2(CO)_8$ and such apical carbon precursors as acetylenes (*140*), dimethyl ketene (*408*), or carbyne chromium complexes (*173*). In several cases (*172,*

220, 305) attempted preparations or thermal rearrangements of compounds with cobalt-silicon bonds resulted in the silicon-containing Co_3C clusters [96]. And the reaction of boron halides and $Co_2(CO)_8$ in most cases leads to the analogous boron-containing Co_3C clusters [97] (*29, 339, 339a*). Aluminium compounds like [97] could also be ob-obtained (*339*), and a special reaction between $BCl_3 \cdot NEt_3$ and $Co_2(CO)_8$ yielded a complex $Co_3(CO)_6(COBCl_2NEt_3)_2$, said to contain two triply-bridging $C(OBCl_2NEt_3)$ ligands (*340*). In these instances the hard acids, silicon and boron, prefer to bind to a carbonyl oxygen atom, thus inducing the formation of the methinyl ligand from a carbonyl group.

$Co_3(CO)_9C{-}OSiR_3$	$Co_3(CO)_9C{-}OBHal_2 \cdot NR_3$
96	*97*

The assumption that the clamp effect of the methinyl ligand stabilizes the Co_3C cluster has been substantiated by bond-energy determinations of $Co_3(CO)_9CHal$ compounds (*187*), which revealed the bond strength order Co–C (methinyl) > Co–C (carbonyl) > Co–Co. The structures of all Co_3C clusters are basically the same. This was shown by a comparison (*317*) of the structures of compounds resulting from pyrolysis of $Co_3(CO)_9CHal$ (*329*), among which are $[Co_3(CO)_9C]_2$ (*54*), $[Co_3(CO)_9C]_2$-CO, $Co_5(CO)_{15}C_3H$, $Co_6(CO)_{18}C_4$, and $Co_8(CO)_{24}C_6$ (*136*), all containing $Co_3(CO)_9C$ units linked by acetylenic carbon chains.

Many ligands substitute CO groups of the basal cobalt triangle in $Co_3(CO)_9CX$. Four isonitrile molecules [*304*] and up to three phosphine or arsine ligands (*289*) can be incorporated. The kinetics of the substitution (*78*) and the fluxional behavior of $(R_3P)Co_3(CO)_8CR$ (*288*) have been investigated. Isomerism in the phosphine-substituted clusters was observed, as well as the frequent occurrence of bridging CO groups (*289*). The ffars ligand with $Co_3(CO)_9CMe$ gives (ffars) $Co_3(CO)_7CMe$ (*119*), whose structure (*155*) shows that two vicinal axial CO groups have been replaced. A maximum of six CO groups are replaced by three $S_2C_2(CF_3)_2$ molecules in $Co_3(CO)_9CX$ to give $L_3Co_3(CO)_3CX$ clusters for which no structure was suggested (*232*).

Arenes replace three CO groups from one basal cobalt atom, as was shown from a crystal structure analysis (*53, 137*), and the arenes can replace one another (*328*). Of the cyclic olefins, norbornadiene replaces two CO groups from one Co to yield a labile complex (*159, 160, 235*), cyclooctatetraene replaces the axial CO ligands from all three cobalt atoms (*53*) and is itself replaced by other Lewis bases (*330*), and cyclopentadiene forms the unusual complex [*98*] with $Co_3(CO)_9CMe$ (*159, 160*). A few catalytic reactions were observed with methinyltricobalt enneacarbonyls including the dimerization of norbornadiene (*160, 235*) and the polymerization of functional olefins (*312*) with different $Co_3(CO)_9CX$.

Considerable advances were made in the organic chemistry of the methinyltricobalt enneacarbonyls. It was shown that the Co_3C core can function as an electron donor as well as an electron acceptor (*143*), allowing electrophilic and nucleophilic attack at the apical carbon atom. Arenes and $Co_3(CO)_9CHal$ form $Co_3(CO)_9CAr$

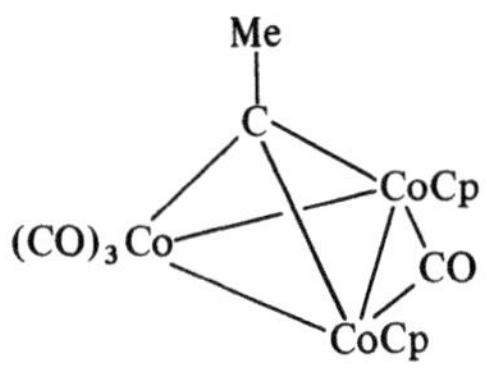

98

under Friedel-Crafts conditions (*144*), while Grignard reagents and $Co_3(CO)_9CHal$ (*145*) or organomercurials and $Co_3(CO)_9CH$ (*348*) lead to $Co_3(CO)_9CR$. The unusual stability of the isolable acylium ion $Co_3(CO)_9C{-}CO^+$ is ascribed to facile charge delocalization. It can be prepared from $Co_3(CO)_9C{-}COOR$ and strong acids (*199, 346*) or from $Co_3(CO)_9CHal$ and $AlCl_3$ (*352*), preferably in the presence of CO (*143*). Its electrophilic character is demonstrated by the ease of its reactions with OH-, NH-, and SH-compounds, with organometallics, reactive arenes, and ferrocene. Electron delocalization is also obvious from the ease of acylation of $Co_3(CO)_9CAr$ clusters (*356*), of formation of the neutral (*345*) and anion (*287*) radicals $Co_3(CO)_9C\cdot$ and $Co_3(CO)_9CX^{\overline{\cdot}}$ with various X substituents.

Another type of compounds stabilized by the Co_3C core are the carbonium ions $Co_3(CO)_9C{-}CR_2^+$. They are isolable intermediates in the Et_3SiH reduction of the ketones $Co_3(CO)_9C{-}COR$ (*349, 354*) and can be obtained by protonation of the alcohols $Co_3(CO)_9C{-}CHOHR$ (*353*) or the olefins $Co_3(CO)_9C{-}CR{=}CHR$ (*344*). They again allow an organic derivative chemistry by reactions with electrophiles (*344, 353*). NMR spectroscopy revealed only small charge differences on the apical and substituent carbon atoms between the neutral $Co_3(CO)_9CR$ and cationic $Co_3(CO)_9C{-}CR_2^+$ clusters, from which conclusions as to the mechanism of charge delocalization could be drawn (*343, 355*). Two other properties of the methinyltricobalt enneacarbonyls probably related to charge distribution are the cobalt–CO (ester) interaction in the compounds $Co_3(CO)_9C{-}CR{=}CR{-}COOR$ (*314*) and the facile CH_2 insertion of $Hg(CH_2Hal)_2$ into the C–H bond of $Co_3(CO)_9CH$ (*351*).

3.7. Rhodium and Iridium

Rhodium is the element of the "big" clusters. Correspondingly, relatively few clusters are known with three and four rhodium atoms, but there is a rich chemistry from hexanuclear complexes upward.

Among the trirhodium systems, the well-known $Cp_3Rh_3(CO)_2$ reacts with trimethylsilyl azide to form $(CpRh)(\mu_3-CO)(\mu_3-NSiMe_3)$ with a structure like [*36*] or [*92*] (*1*). All other new trirhodium clusters contain acetylene ligands and have either the formula $(CpRh)_3(RC{\equiv}CR)$ with the basic structure *99* or $(CpRh)_3$-$(RC{\equiv}CR)(CO)$ with the basic structures [*100*] or [*101*]. They are formed from

$CpRh(CO)_2$ and acetylenes with organic (*188, 224*), fluoroorganic (*138, 139*) or silyl (*326*) substituents or from $Cp(CO)RhJ_2$ and a dilithium acetylide (*189*). The clusters with the triply-bridging CO group [*100*] are fluxional at room temperature, while those with the doubly-bridging CO group [*101*] are rigid (*188, 400*).

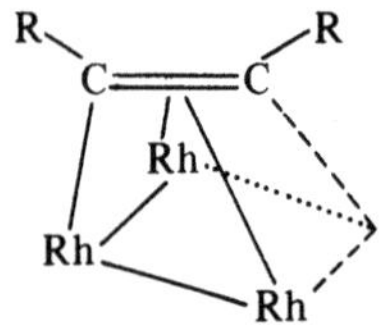

99

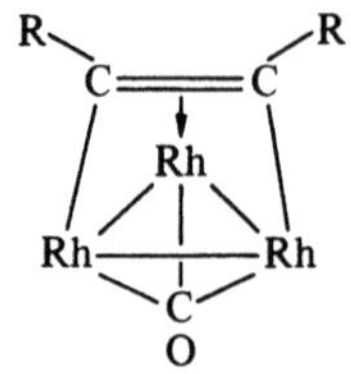

100

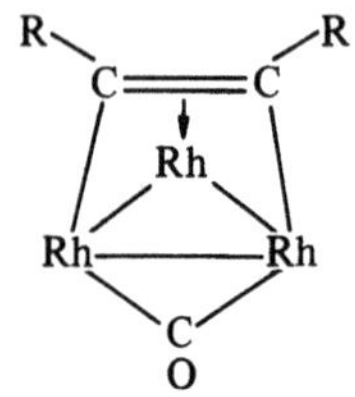

101

Two syntheses of $Rh_4(CO)_{12}$ at ordinary pressure have been developed (*81, 279*) making this compound easily available, and CO scrambling in this molecule has been carefully examined (*110, 112, 164*). Strong nucleophiles including CO at high pressure destroy the Rh_4 cluster (*45, 396*); but with phosphines of limited donor strength, with cyclooctatetraene, and with acetylenes up to four CO groups could be replaced with retention of the Rh_4 unit (*45, 247, 396*). Rh_4-like Rh_6-clusters show catalytic activity (*104, 364*) which has been used for hydroformylation (*88*) and oxidation (*291*) reactions.

The chemistry of $Rh_6(CO)_{16}$ has been explored further. Besides studying the fluxional properties of it and its anion $Rh_6(CO)_{15}^{2-}$ (*203*) prepared from it by reduction (*277*), many substitution reactions were possible. Simple or polymer bound phosphines replace up to six CO groups (*45, 221*), and dienes form $Rh_6(CO)_{14}$ (diene) (*247*). Many anionic groups like Hal, CN, SCN, COOR, CONHR replace one terminal CO ligand (*89*), as was shown by the structure determination of R_4N-$[Rh_6(CO)_{15}J]$ (*11*). Similarly, ethylene in protic solvents forms $Rh_6(CO)_{15}COEt$. Another Rh_6 cluster is trigonal prismatic $Rh_6C(CO)_{15}^{2-}$, which is formed from $Rh_7(CO)_{16}^{3-}$ by incorporation of the carbon atom of a $CHCl_3$ molecule (*13, 22*).

The structures of the larger rhodium clusters have been depicted in Chapter 2.2. $Rh_7(OC)_{16}^{3-}$, which shows intermediate fluxionality (*203*) can be prepared by base disproportionation of $Rh_4(CO)_{12}$ (*22*) or by "construction" from $Rh_6(CO)_{16}$ and $Rh(CO)_4^-$ (*277*). Several reactions starting from $Rh_6(CO)_{16}$ lead to bromide- or iodide-substituted $Rh_7(CO)_{16}Hal^{2-}$ (*280*) with structure [*13*] (*19*). Oxidation of $Rh_6C(CO)_{15}^{2-}$ under CO yields another carbido-cluster $Rh_8C(CO)_{19}$ (*14*) with structure [*14*] (*23*). The Rh–Rh bonded Rh_6 dimer $Rh_{12}(CO)_{30}^{2-}$, of which extensive CO scrambling might be expected, is rigid at all temperatures (*90*). Its hydrogenation at 80° results in the formation of $Rh_{13}(CO)_{24}H_2^{3-}$ or $Rh_{13}(CO)_{24}H_3^{2-}$, the first clusters which have a structure [*17*] corresponding to the closely packed structures of pure metals (*12*). And the biggest rhodium cluster with a proved structure [*18*] is $Rh_{15}C_2(CO)_{28}^{2-}$, obtained by oxidation of $Rh_6C(CO)_{15}^{2-}$ under CO (*14*).

Heterometallic rhodium clusters exist with iron, cobalt, and iridium (see below). The structures [*102*] and [*103*] were suggested from low-temperature NMR studies for the trinuclear Rh(Cp)–Fe(CO) clusters (*212*), and the crystal structures of two tetranuclear Rh(Cp)–Fe(CO) clusters [*104*] (*98*) and [*105*] (*97*) were determined. The mixed cobalt-rhodium clusters of the $M_4(CO)_{12}$ type $Co_3Rh(CO)_{12}$ and $Co_2Rh_2(CO)_{12}$ could be obtained in several ways starting from cationic and anionic derivatives of both metals (*278*), or by intermolecular ligand exchange between $Co_2Rh_2(CO)_{12}$ and $Co_2Rh_2(CO)_8(PF_3)_4$, leading to $Co_2Rh_2(CO)_{10}(PF_3)_2$ (*208*). The different degree of ease of CO scrambling at the cobalt or rhodium atoms in $Co_3Rh(CO)_{12}$ was shown by an NMR study (*228*). Hexanuclear cobalt-rhodium clusters are $Co_2Rh_4(CO)_{16}$, obtained by pyrolysis of $Co_2Rh_2(CO)_{12}$ (*278*), and $Co_2Rh_4C(CO)_{15}^{2-}$, obtained from $Co_3(CO)_9CCl$ and $Rh(CO)_4^-$ (*14*).

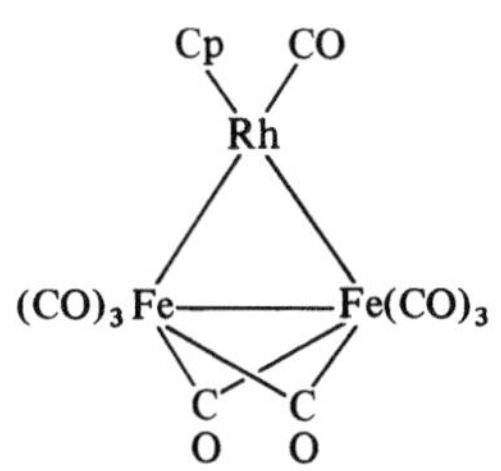

102

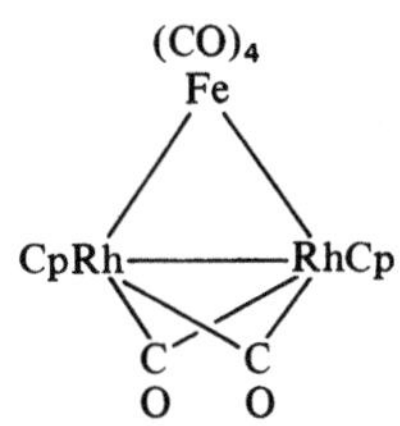

103

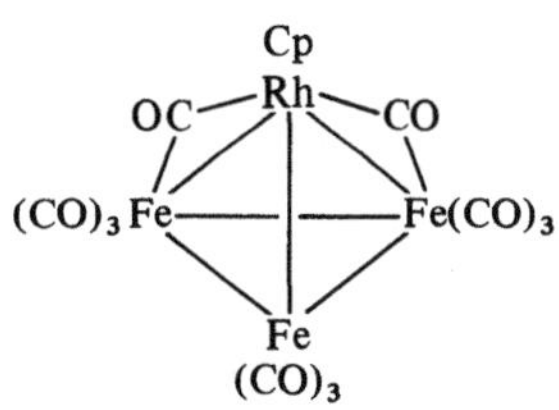

104

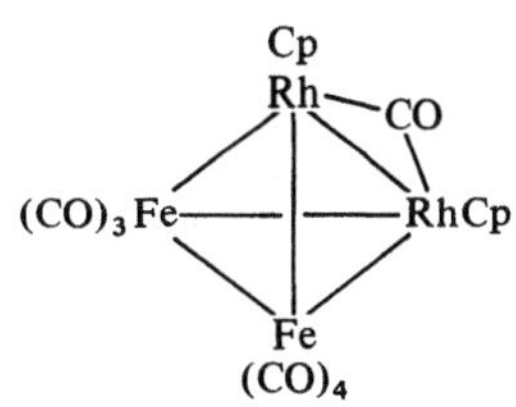

105

Of the trinuclear iridium clusters, only the acetylene-bridged $(CpIr)_3(RC{\equiv}CR)$ of structure [*99*] and $(CpIr)_3(RC{\equiv}CR)(CO)$ of structure [*101*] were described (*188*). Three new syntheses for $Ir_4(CO)_{12}$ were reported (*275, 369, 397*) whose two-step process (*369*) gives 85% yield. Phosphines replace up to four CO groups in $Ir_4(CO)_{12}$ (one per iridium atom) (*80, 147, 238, 395*); the cooperative nature of this process (*238*) has been mentioned already. Strong acids protonate $Ir_4(CO)_8(PR_3)_4$ (*80*), and strong bases including $Ir(CO)_4^{2-}$ and reducing agents produce the new clusters

$HIr_4(CO)_{11}^-$, $Ir_6(CO)_{15}^{2-}$, $Ir_6(CO)_{16}$, $Ir_8(CO)_{20}^{2-}$, and $Ir_8(CO)_{22}^{2-}$, of which $Ir_6(CO)_{15}^{2-}$ and $Ir_6(CO)_{16}$ are isostructural to the corresponding cobalt and rhodium compounds (*28*). The mixed clusters, $Co_2Ir_2(CO)_{12}$ and $Rh_2Ir_2(CO)_{12}$, were prepared from $Ir(CO)_2J_2^-$ and $Co(CO)_4^-$ and $Rh(CO)_2Cl_2^-$ and $Ir(CO)_4^-$ (*278*). The structure of $Co_2Ir_4(CO)_{12}$ is like that of $Co_4(CO)_{12}$ and unlike that of $Ir_4(CO)_{12}$ (*21*).

3.8. Nickel, Palladium, and Platinum

Nickel forms organometallic clusters with three to six metal atoms. Among these some unusual structures and bonding situations occur which were mentioned in Chapter 2. The cluster chemistry of palladium is rather poor, and the outstanding features of platinum are a considerable number of trinuclear heterometallic complexes and the chimney-like structures of the clusters $[Pt_3(CO)_6]_n^{2-}$.

The trinuclear nickel clusters reported contain the $(CpNi)_3$ unit together with triply-bridging ligands. Of the $Cp_3Ni_3(NR)$ class the synthesis of the phenyl and *n*-butyl compound was reported (*300*), and the crystal structure of the *t*-butyl compound was determined (*236*). The compound Cp_3Ni_3CPh, obtained from $NiCp_2$ and $PhCH_2MgCl$ (*388*), is isoelectronic to the methinyltricobalt enneacarbonyls. And the long-known $Cp_3Ni_3(CO)_2$ adds boron halides at the triply-bridging carbonyl oxygen (*266*).

An unusual tetrahedral Ni_4-cluster with three unpaired electrons and three face-bridging hydrogen atoms is $Cp_4Ni_4H_3$ (*218*), prepared from CpNiNO and $NaBH_4$ (*299*). Even more unusually, face-bridging ligands are present in $Ni_4(CO)_4$-$(CF_3C{\equiv}CCF_3)_3$ (*120*) and $Ni_4(CNBu^t)_7$ (*122*), which are both of interest with respect to catalytic reactions because of their fluxionality and lability. The pentanickel cluster $Ni_5(CO)_{12}^{2-}$ is obtained together with the hexanickel cluster $Ni_6(CO)_{12}^{2-}$ by reduction or base disproportionation of $Ni(CO)_4$ (*274*). $Ni_5(CO)_{12}^{2-}$ has the rare trigonal bipyramidal structure [*10*] (*274*), whereas $Ni_6(CO)_{12}^{2-}$ is a distorted 86-electron octahedral system (*71*), thereby differing markedly from trigonal prismatic $Pt_6(CO)_{12}^{2-}$ of structure [*11*].

The composition and structure of the molybdenum- and tungsten-containing nickel clusters $M_2Ni_3(CO)_{16}^{2-}$ (*333*), obtained from $Ni(CO)_4$ and $M_2(CO)_{10}^{2-}$, are closely related to that of $Ni_5(CO)_{12}^{2-}$ containing a basal $Ni_3(CO)_6$ triangle. The related cluster $Mo_2Ni_4(CO)_{14}^{2-}$ differs from those in that it has a basal $Ni_4(CO)_6$ rhombus (*333*). The reactions of $Co_2(CO)_8$, $Fe_2(CO)_9$, and $Mn(CO)_5^-$ with $[CpNi(CO)]_2$ lead to $CpNiCo_3(CO)_9$ with structure [*106*] and to $Cp_2Ni_2Fe(CO)_5$ and $Cp_2Ni_2Mn(CO)_5^-$ with structure [*107*] (*213*). In $CpNiCo_3(CO)_9$, one CO ligand can be replaced by isonitriles (*304*). Other cobalt-nickel clusters are $NiCo_3(CO)_{11}^-$, of unknown structure obtained from $NiCl_2$ and $Co[Co(CO)_4]_2$, and its thermolysis product $Ni_2Co_4(CO)_{14}^{2-}$ (*87*), which is isoelectronic and structurally related (*20*) to $Co_6(CO)_{14}^{4-}$.

Only three palladium clusters were reported. Sulfur dioxide reacts with $Pd(CNBu^t)_2$ to form $Pd_3(SO_2)_2(CNBu^t)_5$ (*310*) of structure [*108*] (*309*), and reduc-

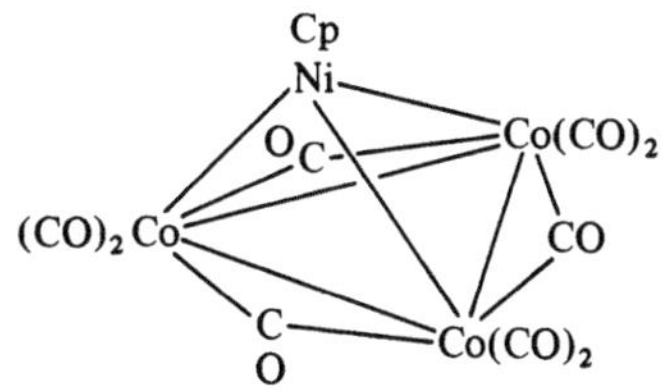

106

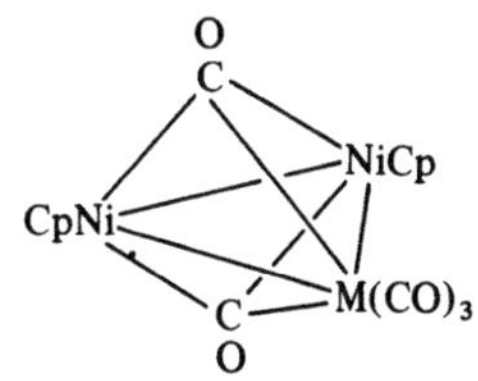

107, M = Fe,Mn

tions of triphenyl phosphine palladium complexes in the presence of CO yield $Pd_3(CO)_3(PPh_3)_3$ and $Pd_3(CO)_3(PPh_3)_4$, for which the structures [*109*] and [*110*] were suggested by analogy with the corresponding platinum compounds (*205, 267*).

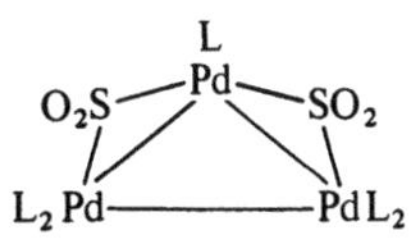

108, L = CNBu^t

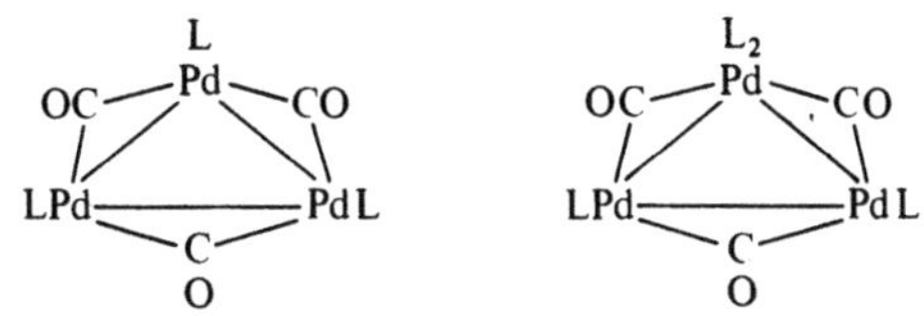

109 L = PPh_3 *110*

In addition to the known platinum clusters $Pt_3(CO)_3(PPh_3)_n$ ($n = 3, 4$),(*399*) another compound, $Pt_3(CO)_2(PPh_3)_4$, could be obtained for which structure [*111*] was suggested (*52*). Other triangular platinum clusters include $Pt_3(CNBu^t)_6$, obtained from $Pt(COD)_2$ and $CNBu^t$ of structure [*112*] (*193*), $Pt_3(PPh_3)_2(PPh_2)_3Ph$ obtained from $Pt(PPh_3)_4$ by pyrolysis of structure [*113*] (*372*), and the complexes $Pt_3(PPh_3)_4(CO)$-$(CPh{=}CHMe)^+$ and $Pt_3(PMePh_2)_5(CO)(CPh{=}CPhH)^+$, obtained from phosphine-substituted platinum compounds with acetylenes, CO, and acids (*44*).

The beautiful series of clusters $Pt_6(CO)_{12}^{2-}$, $Pt_9(CO)_{18}^{2-}$, and $Pt_{15}(CO)_{30}^{2-}$, with the structures [*11*], [*19*], and [*20*] is one of the rare examples where a systematic sequence is present in the preparations as well as in the chimney-like structures (*72*).

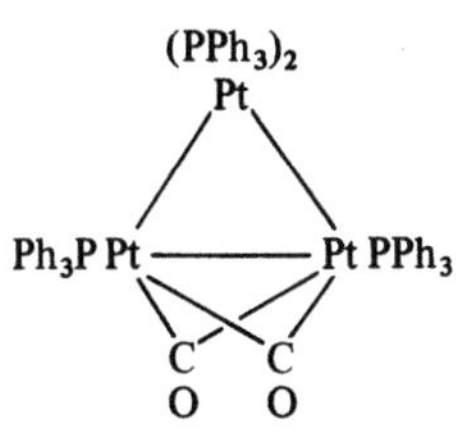

111

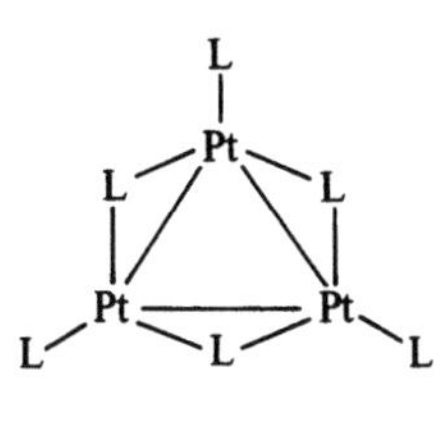

112, L = CNBu^t

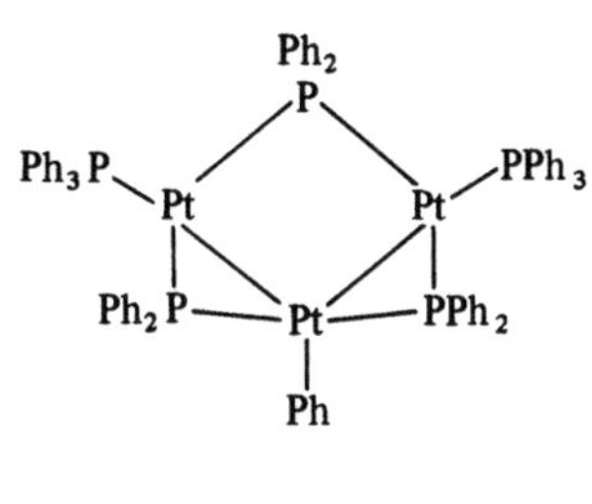

113

Depending on the amount of NaOH used, the reduction of Na_2PtCl_6 yields the clusters with 15, 12, 9, 6, or 3 platinum atoms.

The highest number of platinum clusters are of the hetero-metallic kind. Chelate phosphine ligand complexes L_2PtCl_2 react with $Co(CO)_4^-$ to form triangular $L_2PtCo_2(CO)_7$ (*134*). Phosphine complexes PtL_4 react with iron, ruthenium, and osmium carbonyls to form a series of heterotrimetallic clusters of the general formulas $PtM_2(CO)_9L$, $PtM_2(CO)_8L_2$, $PtM_2(CO)_7L_3$, $MPt_2(CO)_5L_3$, and $MPt_2(CO)_4L_4$ (*16, 62, 63, 64*). For these the structures of [*114*] (*285, 286*), [*115*] (*295*), and [*116*] (*15*) were determined.

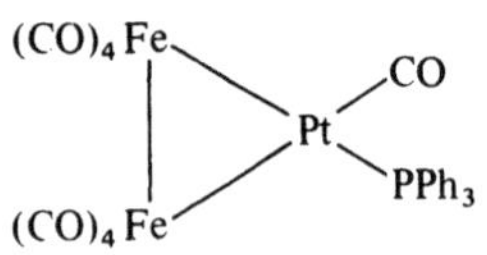

114

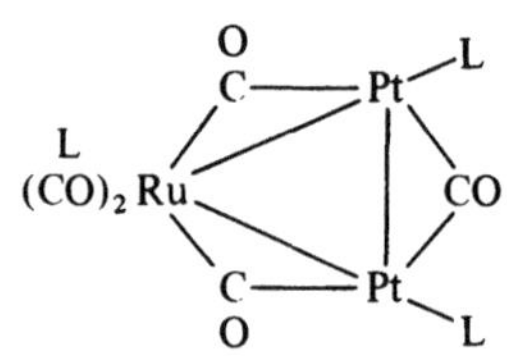

115, L = $PMePh_2$

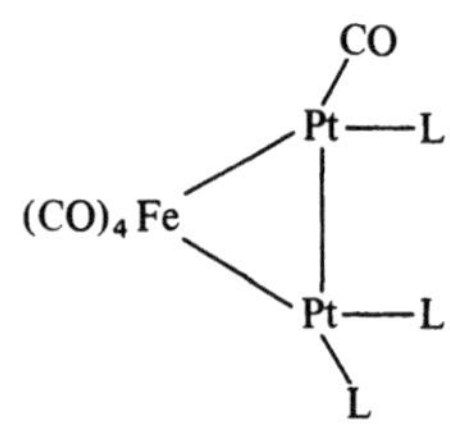

116, L = $P(OPh)_3$

Tetranuclear heterometallic platinum clusters adopt the butterfly structure exemplified by $Co_2Pt_2(CO)_8(PPh_3)_2$ (structure [*2*]) (*174*), obtained from $(PPh_3)_2PtCl_2$ and $Co(CO)_4^-$ (*52*). Similar compounds, but supposedly with a Pt–Pt bond in the center of the butterfly, are $Mo_2Pt_2Cp_2(PPh_3)_2(CO)_6$, obtained from $(PPh_3)_2PtCl_2$ and $MoCp(CO)_3^-$ (*52*), and $H_2Os_2Pt_2(PPh_3)_2(CO)_8$, obtained from $H_2Os(CO)_4$ and $Pt(PPh_3)_4$ (*64*).

3.9. Copper, Silver, and Gold

Organo-copper cluster chemistry has only been developed in recent years as a result of the discovery of the effect of "internal" ligands (*306*). The few examples of organo-silver clusters resemble their copper counterparts, while the cluster chemistry of gold, though limited, is unique.

The stabilizing effect of internal ligands on arylcopper tetramers is demonstrated by structure [*3*] (*194, 263*). Whereas simple arylcoppers are polymeric, the use of aryl groups with substituents containing NMe_2, OMe, PPh_2, SO_2NMe_2, and F substituents (*69, 256–258*) blocks coordination sites on copper and lowers the nuclearity to four. Addition of further nucleophiles, such as isonitriles or phosphines, breaks the clusters up into dinuclear or mononuclear complexes (*259, 260*). The unsaturated nature of the tetramers is further demonstrated by the ligand exchange between $[CuAr]_4$ and $[CuAr']_4$, which probably goes via octanuclear intermediates (*263*), and by the addition of copper halides to oligomeric CuAr · CuHal species (*256, 264*) and of arylmagnesiums to Cu_4MgAr_6 (*342*). There are only a few tetrameric organocoppers which do not belong to the internal ligand class. To these belong $[Cu-CH_2SiMe_3]_4$, of structure [*4*] (*225*) and prepared from CuJ and $LiCH_2SiMe_3$ (*269*), and the more inorganic cluster $Cu_4(S_2PR_2)_4$ (*270*).

Hexanuclear copper clusters with internal ligands can be obtained from $[Cu-C_6H_4-o-NMe_2]_4$ and CuHal (*256*). They have the stoichiometry $(CuAr)_4$-$(CuHal)_2$ and contain an octahedron of copper atoms with two edge-bridging halogen atoms (*195, 265*). Their reaction with organolithium reagents leads to $Ar_4R_2Cu_6$ with a similar structure (*261*). The compound $[HCu(PPh_3)]_6$ prepared from $[ClCu(PPh_3)]_4$ and $NaBH(OMe)_3$ is a rare example of an 84-electron octahedral cluster (*42, 92*). Another octahedral complex is $Cu_4Ir_2(PPh_3)_2(C{\equiv}CPh)_8$, obtained from $IrCl(CO)(PPh_3)_2$ and $CuC{\equiv}CPh$ (*8*), the structure of which (*91*) consists of a ligand-free Cu_4 square capped and isolated by two $Ir(PPh_3)(C{\equiv}CPh)_4$ units.

Two octanuclear copper complexes were reported. Of these, the arrangement of the copper atoms in a cube was suggested for $[Cu-C_6H_4-m-CF_3]_8$ (*70*) and proved for the anion $Cu_8(dts)_6^{4-}$ (dts = dithiosquarate) (*206*).

Tetrameric silver clusters stabilized by internal ligands from NMe_2- and OMe-substituted aryl groups seem to exist, too (*272*). Silver can also be incorporated into hexanuclear clusters of the type $Cu_{6-n}Ag_nAr_4Br_2$ with $n = 0, 2, 4, 6$ by reactions of $[MAr]_4$ with MBr_2 (M = Cu, Ag) in the corresponding ratio (*256, 265, 272*).

Gold clusters do not follow the structural trends observed for copper and silver; and internal ligand arylgold tetramers could only be obtained as the complexes $Cu_{4-n}Au_nAr_4$ containing copper atoms (*262*). Instead, gold forms a few clusters not predictable in terms of electron counting or structural systematics. The triphenylphosphorin-AuCl adduct reacts with methanol to a cluster $[Au-LOMe]_3$ (L = triphenylphosphorin) formulated with a gold triangle (*237*). The very unusual situation of a 76-electron octahedron is realized in the cation $Au_6(PAr_3)_6^{2+}$, obtained from $Ar_3P \cdot AuNO_3$ and $NaBH_4$ (*38, 40*). More unusual, however, are the clusters $Au_9(PR_3)_8X_3$ and $Au_{11}(PR_3)_7X_3$, which are also obtained from the $R_3P \cdot AuX$ compounds by borohydride reduction (*37, 77*). Their structures [*15*] (*37*) and [*16*] (*39*), which can be represented by the basic formulas $AuAu_8$ and $AuAu_{10}$, bear no resemblance to simple polyhedra nor to the atom arrangements in plain metals. And electron counting as discussed in Chapter 2.3. does not lead to an obvious description of their bonding situation. It is amazing that such a wealth of polynuclear structures of the group Ib elements should be followed by the complete absence of cluster chemistry of the group IIb elements.

4. Concluding Remarks

Considerable advances in the field of transition metal cluster chemistry have been made during the last five years. They have confirmed that in many respects a cluster complex is comparable to a metallic surface. They have also shown that clusters allow reactions which are not observed with simple metal complexes. And they have finally demonstrated that structural and bonding properties of clusters require new concepts for their description.

It cannot be stated, however, that a systematic cluster chemistry exists yet. In most cases each metal has its own structural and chemical cluster properties, and only a few interrelations exist. Predictability is low for cluster compounds and reactions. This is underlined by the fact that there is still no systematic way of preparing clusters. And the few generalizations that can be drawn from the large number of observations justify the conclusion that this field of chemistry is still in its beginning stages. In order to keep cluster research free from the odor of being a *l'art pour l'art* science, considerable efforts toward a better understanding will have to be made.

References

1. *Abel, E. W., Blackmore, T., Whitley, R. J.:* Inorg. Nucl. Chem. Letters *10*, 941 (1974).
2. *Abel, E. W., Crosse, B. C.:* Organomet. Chem. Rev. *2*, 443 (1967).
3. *Abel, E. W., Stone, F. G. A.:* Quart. Rev. Chem. Soc. *23*, 325 (1969).
4. *Abel, E. W., Stone, F. G. A.:* Quart. Rev. Chem. Soc. *24*, 498 (1970).
5. *Abel, E. W., Stone, F. G. A.:* Specialist Periodical Reports, Organometallic Chemistry *1*, 184 (1972); *2*, 224 (1973), *3*, 169 (1974); *4*, 162 (1975).
6. *Abu Salah, O. M., Bruce, M. I.:* J.C.S. Dalton *1974*, 2302.
7. *Abu Salah, O. M., Bruce, M. I.:* J.C.S. Dalton *1975*, 2311.
8. *Abu Salah, O. M., Bruce, M. I., Churchill, M. R., Bezman, S. A.:* Chem. Commun. *1972*, 858.
9. *Aime, S., Gambino, O., Milone, L., Sappa, E., Rosenberg, E.:* Inorg. Chim. Acta *15*, 53 (1975).
10. *Aime, S., Milone, L., Sappa, E.:* Inorg. Chim. Acta *16*, L 7 (1975).
11. *Albano, V. G., Bellon, P. L., Sansoni, M.:* J. Chem. Soc. *A 1971*, 678.
12. *Albano, V. G., Ceriotti, A., Chini, P., Ciani, G., Martinengo, S., Anker, W. M.:* Chem. Commun. *1975*, 859.
13. *Albano, V. G., Chini, P., Martinengo, S., McCaffrey, D. J. A., Strumolo, D., Heaton, B. T.:* J. Amer. Chem. Soc. *96*, 8106 (1974).
14. *Albano, V. G., Chini, P., Martinengo, S., Sansoni, M., Strumolo, D.:* Chem. Commun. *1974*, 299.
15. *Albano, V. G., Ciani, G.:* J. Organomet. Chem. *66*, 311 (1974).
16. *Albano, V. G., Ciani, G.; Bruce, M. I., Shaw, G., Stone, F. G. A.:* J. Organomet. Chem. *42*, C 99 (1972).
17. *Albano, V. G., Ciani, G., Freni, M., Romiti, P.:* J. Organomet. Chem. *96*, 259 (1975).
18. *Albano, V. G., Ciani, G., Manassero, M., Sansoni, M.:* J. Organomet. Chem. *34*, 353 (1972).
19. *Albano, V. G., Ciani, G., Martinengo, S., Chini, P., Giordano, G.:* J. Organomet. Chem. *88*, 381 (1975).
20. *Albano, V. G., Ciano, G., Chini, P.:* J.C.S. Dalton *1974*, 432.
21. *Albano, V. G., Ciano, G., Martinengo, S.:* J. Organomet. Chem. *78*, 265 (1974).
22. *Albano, V. G., Sansoni, M., Chini, P., Martinengo, S.:* J.C.S. Dalton *1973*, 651.
23. *Albano, V. G., Sansoni, M., Chini, P., Martinengo, S., Strumolo, D.:* J.C.S. Dalton, *1975*, 305.
24. *Albini, A., Kisch, H.:* J. Organomet. Chem. *94*, 75 (1975).
25. *Albini, A., Kisch, H.:* Angew. Chem. *87*, 206 (1975), Angew. Chem. Int. Ed. Engl. *14*, 182 (1975).
26. *Alper, H.:* Inorg. Chem. *11*, 976 (1972).
27. *Alper, H.; Chan, H. S. K.:* Inorg. Chem. *13*, 225 (1974).
28. *Angoletta, M., Malatesta, C., Caglio, G.:* J. Organomet. Chem. *94*, 99 (1975).
29. *Bätzel, V., Müller, U., Allmann, R.:* J. Organomet. Chem. *102*, 109, (1975).
30. *Baird, M. C.:* Progr. Inorg. Chem. *9*, 1 (1968).
31. *Barnett, B. L., Krüger, C.:* Angew. Chem. *83*, 969 (1971), Angew. Chem. Int. Ed. Engl. *10*, 910 (1971).
32. *Bau, R., Burt, J. C., Knox, S. A. R., Laine, R. M., Philips, R. P., Stone, F. G. A.:* Chem. Commun. *1973*, 726.
33. *Bau, R., Chou, B. C. K., Knox, S. A. R., Riera, V., Stone, F. G. A.:* J. Organomet. Chem. *82*, C 43 (1974).
34. *Bau, R., Don, B., Greatrex, R., Haines, R. J., Love, R. A., Wilson, R. D.:* Inorg. Chem. *14*, 3021 (1975).
35. *Beck, W., Danzer, W., Thiel, G.:* Angew. Chem. *85*, 625 (1973), Angew. Chem. Int. Ed. Engl. *12*, 582 (1973).
36. *Belford, R., Taylor, H. D., Woodward, P.:* J.C.S. Dalton *1972*, 2425.
37. *Bellon, P. L., Cariati, F., Manassero, M., Naldini, C., Sansoni, M.:* Chem. Commun. *1971*, 1423.

38. *Bellon, P.L., Manassero, M., Naldini, L., Sansoni, M.:* Chem. Commun. *1972*, 1035.
39. *Bellon, P.L., Manassero, M., Sansoni, M.:* J.C.S. Dalton *1972*, 1481.
40. *Bellon, P., Manassero, M., Sansoni, M.:* J.C.S. Dalton, *1973*, 2423.
41. *Bernal, I., Davis, B.R., Good, M.L., Chandra, S.:* J. Coord. Chem. *2*, 61 (1972).
42. *Bezman, S.A., Churchill, M.R., Osborn, J.A., Wormold, J.:* J. Amer. Chem. Soc. *93*, 2063 (1971).
43. *Bird, P.H., Fraser, A.R.:* J. Organomet. Chem. *73*, 103 (1974).
44. *Blake, D.M., Leung, L.M.:* Inorg. Chem. *11*, 2879 (1972).
45. *Booth, B.L., Else, M.J., Fields, R., Haszeldine, R.N.:* J. Organomet. Chem. *27*, 119 (1971).
46. *Booth, B.L., Haszeldine, R.N., Inglis, T.:* J.C.S. Dalton *1975*, 1449.
47. *Bor, G., Sbrignadello, G., Marcati, F.:* J. Organomet. Chem. *46*, 357 (1972).
48. *Braddock, J.N., Meyer, T.J.:* Inorg. Chem. *12*, 723 (1973).
49. *Bradford, C.W., Nyholm, R.S.:* J. Chem. Soc. *A 1971*, 2038.
50. *Bradford, C.W., Nyholm, R.S.:* J.C.S. Dalton *1973*, 529.
51. *Bradford, C.W., Nyholm, R.S., Gainsford, G.J., Guss, J.M., Ireland, P.R., Mason, R.:* Chem. Commun. *1972*, 87.
52. *Braunstein, P., Dehand, J., Nennig, J.F.:* J. Organomet. Chem. *92*, 117 (1975).
53. *Brice, M.D., Dellaca, R.J., Penfold, B.R., Spencer, J.L.:* Chem. Commun. *1971*, 72.
54. *Brice, M.D., Penfold, B.R.:* Inorg. Chem. *11*, 1381 (1972).
55. *Brookes, A., Knox, S.A.R., Stone, F.G.A.:* J. Chem. Soc. *A 1971*, 3469.
56. *Bruce, M.I.:* J. Organomet. Chem. *48*, 303 (1973), *53*, 141 (1973), *58*, 153 (1973), *75*, 335 (1974), *89*, 215 (1975).
57. *Bruce, M.I., Abu Salah, O.M., Davis, R.E., Raghavan, N.V.:* J. Organomet. Chem. *64*, C48 (1974).
58. *Bruce, M.I., Bennett, R.L., Stone, F.G.A.:* J. Organomet. Chem. *38*, 325 (1972).
59. *Bruce, M.I., Cairns, M.A., Green, M.:* J.C.S. Dalton *1972*, 1293.
60. *Bruce, M.I., Goodall, B.L., Sheppard, G.L., Stone, F.G.A.:* J.C.S. Dalton *1975*, 591.
61. *Bruce, M.I., Iqbal, M.Z., Stone, F.G.A.:* J. Organomet. Chem. *31*, 275 (1971).
62. *Bruce, M.I., Shaw, G., Stone, F.G.A.:* Chem. Commun. *1971*, 1288.
63. *Bruce, M.I., Shaw, G., Stone, F.G.A.:* J.C.S. Dalton *1972*, 1082.
64. *Bruce, M.I., Shaw, G., Stone, F.G.A.:* J.C.S. Dalton *1972*, 1781.
65. *Bruce, M.I., Shaw, G., Stone, F.G.A.:* J.C.S. Dalton *1972*, 2094.
66. *Bruce, M.I., Shaw, G., Stone, F.G.A.:* J.C.S. Dalton *1973*, 1667.
67. *Buckingham, A.D., Yesinowski, J.P., Canty, A.J., Rest, A.J.:* J. Amer. Chem. Soc. *95*, 2732 (1973).
68. *Burt, J.C., Knox, S.A.R., Stone, F.G.A.:* J.C.S. Dalton *1975*, 73.
69. *Cairncross, A., Omura, H., Sheppard, W.A.:* J. Amer. Chem. Soc. *93*, 248 (1971).
70. *Cairncross, A., Sheppard, W.A.:* J. Amer. Chem. Soc. *93*, 247 (1971).
71. *Calabrese, J.C., Dahl, L.F., Cavalieri, A., Chini, P., Longoni, G., Martinengo, S.:* J. Amer. Chem. Soc. *96*, 2616 (1974).
72. *Calabrese, J.C., Dahl, L.F., Chini, P., Longoni, G., Martinengo, S.:* J. Amer. Chem. Soc. *96*, 2614 (1974).
73. *Camus, A., Nardin, G., Randaccio, L.:* Inorg. Chim. Acta *12*, 23.
74. *Canty, A.J., Domingos, A.J.P., Johnson, B.F.G., Lewis, J.:* J.C.S. Dalton *1973*, 2056.
75. *Canty, A.J., Johnson, B.F.G., Lewis, J.:* J. Organomet. Chem. *43*, C 35 (1972).
76. *Canty, A.J., Johnson, B.F.G., Lewis, J., Norton, J.R.:* Chem. Commun. *1972*, 1331.
77. *Cariati, F., Naldini, L.:* Inorg. Chim. Acta *5*, 172 (1971).
78. *Cartner, A., Cunninghame, R.G., Robinson, B.H.:* J. Organomet. Chem. *92*, 49 (1975).
79. *Carty, A.J., Ferguson, G., Paik, H.N., Restivo, R.:* J. Organomet. Chem. *74*, C 14 (1974).
80. *Cattermole, P.E., Orrell, K.G., Osborne, A.G.:* J.C.S. Dalton *1974*, 328.
81. *Cattermole, P.E., Osborne, A.G.:* J. Organomet. Chem. *37*, C 17 (1972).
82. *Chaudhuri, M.K., Haas, A., Welcman, N.:* J. Organomet. Chem. *85*, 85 (1975).
83. *Chaudhuri, M.K., Haas, A., Welcman, N.:* J. Organomet. Chem. *91*, 81 (1975).
84. *Chia, L.S., Cullen, W.R., Sams, J.R.: Ward, J.E.H.:* Can. J. Chem. *51*, 3223 (1973).

85. *Chini, P.:* Inorg. Chim. Acta Rev. *2*, 31 (1968).
86. *Chini, P.:* Pure Appl. Chem. *23*, 489 (1970).
87. *Chini, P., Cavalieri, A., Martinengo, S.:* Coord. Chem. Rev. *8*, 3 (1972).
88. *Chini, P., Martinengo, S., Garlaschelli, G.:* Chem. Commun. *1972*, 709.
89. *Chini, P., Martinengo, S., Giordano, G.:* Gazzetta *102*, 330 (1972).
90. *Chini, P., Martinengo, S., McCaffrey, D.J.A., Heaton, B.T.:* Chem. Commun. *1974*, 310, 804.
91. *Churchill, M.R., Bezman, S.A.:* Inorg. Chem. *13*, 1418 (1974).
92. *Churchill, M.R., Bezman, S.A., Osborne, J.A., Wormald, J.:* Inorg. Chem. *11*, 1818 (1972).
93. *Churchill, M.R., Chang, S.W.Y.:* Chem. Commun. *1974*, 248.
94. *Churchill, M.R., DeBoer, B.G.:* Inorg. Chem. *14*, 2502 (1975).
95. *Churchill, M.R., DeBoer, B.G., Rotella, F.J., Abel, E.W., Rowley, R.J.:* J. Amer. Chem. Soc. *97*, 7158 (1975).
96. *Churchill, M.R., Scholar, F.R., Wormald, J.:* J. Organomet. Chem. *28*, C 21 (1971).
97. *Churchill, M.R., Veidis, M.V.:* J. Chem. Soc. *A 1971*, 2170.
98. *Churchill, M.R., Veidis, M.V.:* J. Chem. Soc. *A 1971*, 2995.
99. *Churchill, M.R., Wormald, J.:* J. Amer. Chem. Soc. *93*, 5670 (1971).
100. *Churchill, M.R., Wormald, J.:* Inorg. Chem. *12*, 191 (1973).
101. *Churchill, M.R., Wormald, J.:* J.C.S. Dalton *1974*, 2410.
102. *Churchill, M.R., Wormald, J., Knight, J., Mays, M.J.:* J. Amer. Chem. Soc. *93*, 3073 (1971).
103. *Cohen, M.A., Kidd, D.R., Brown, T.L.:* J. Amer. Chem. Soc. *97*, 4408, (1975).
104. *Collman, J.P., Hegedus, L.S., Cooke, M.P., Norton, J.R., Dolcetti, G., Marquardt, D.N.:* J. Amer. Chem. Soc. *94*, 1789 (1972).
105. *Connelly, N.G.:* Specialist Periodical Reports, Organometallic Chemistry *2*, 234 (1973), *3*, 180 (1974), *4*, 172 (1975).
106. *Cooke, C.G., Mays, J.M.:* J. Organomet. Chem. *74*, 449 (1974).
107. *Cooke, C.G., Mays, M.J.:* J.C.S. Dalton *1975*, 455.
108. *Cooke, C.G., Mays, M.J.:* J. Organomet. Chem. *88*, 231 (1975).
109. *Cotton, F.A.:* Quart. Rev. Chem. Soc. *20*, 389 (1966).
109a. *Cotton, F.A.:* Acc. Chem. Res. *1*, 257 (1968).
110. *Cotton, F.A.:* Bull. Soc. Chim. France *1973*, 2587.
111. *Cotton, F.A., Hunter, D.L.:* Inorg. Chim. Acta *11*, L9 (1974).
112. *Cotton, F.A., Kruczynski, L., Shapiro, B.L., Johnson, L.F.:* J. Amer. Chem. Soc. *94*, 6191 (1972).
113. *Cotton, F.A., Troup, J.M.:* J. Amer. Chem. Soc. *96*, 4155 (1974).
114. *Cotton, F.A., Troup, J.M.:* J. Amer. Chem. Soc. *96*, 5070 (1974).
115. *Cotton, J.D.:* Specialist Periodical Reports, Organometallic Chemistry, *1*, 194 (1972).
116. *Cowley, A.H., Hill, K.E.:* Inorg. Chem. *12*, 1446 (1973).
117. *Cox, A., Woodward, P.:* J. Chem. Soc. *A 1971*, 3599.
118. *Crow, J.P., Cullen, W.R.:* Inorg. Chem. *10*, 1529 (1971).
119. *Crow, J.P., Cullen, W.R.:* Inorg. Chem. *10*, 2165 (1971).
120. *Davidson, J.L., Green, M., Stone, F.G.A., Welch, A.J.:* J. Amer. Chem. Soc. *97*, 7490 (1975).
121. *Dawes, J.L., Holmes, J.D.:* Inorg. Nucl. Chem. Letters 7, 847 (1971).
122. *Day, V.W., Day, R.O., Kristoff, J.S., Hirsekorn, F.J., Muetterties, E.L.:* J. Amer. Chem. Soc. *97*, 2571 (1975).
123. *Deeming, A.J., Ettorre, R., Johnson, B.F.G., Lewis, J.:* J. Chem. Soc. *A 1971*, 1797, 2701.
124. *Deeming, A.J., Hasso, S.:* J. Organomet. Chem. *88*, C 21 (1975).
125. *Deeming, A.J., Hasso, S., Underhill, M.:* J. Organomet. Chem. *80*, C 53 (1974).
126. *Deeming, A.J., Hasso, S., Underhill, M.:* J.C.S. Dalton *1975*, 1614.
127. *Deeming, A.J., Hasso, S., Underhill, M., Canty, A.J., Johnson, B.F.G., Jackson, W.G., Lewis, J., Matheson, T.W.:* Chem. Commun. *1974*, 807.
128. *Deeming, A.J., Kimber, R.E., Underhill, M.:* J.C.S. Dalton *1973*, 2589.

129. *Deeming, A.J., Nyholm, R.S., Underhill, M.:* Chem. Commun. *1972*, 224.
130. *Deeming, A.J., Underhill, M.:* J. Organomet. Chem. *42*, C 60 (1972).
131. *Deeming, A.J., Underhill, M.:* Chem. Commun. *1973*, 277.
132. *Deeming, A.J., Underhill, M.:* J.C.S. Dalton *1973*, 2727.
133. *Deeming, A.J., Underhill, M.:* J.C.S. Dalton *1974*, 1415.
134. *Dehand, J., Nennig, J.F.:* Inorg. Nucl. Chem. Letters *10*, 875 (1974).
135. *Delbaere, L.T.J., Kruczynski, L.J., McBride, D.W.:* J.C.S. Dalton *1973*, 307.
136. *Dellaca, R.J., Penfold, B.R.:* Inorg. Chem. *10*, 1269 (1971).
137. *Dellaca, R.J., Penfold, B.R.:* Inorg. Chem. *11*, 1855 (1972).
138. *Dickson, R.S., Kirsch, H.P.:* Aust. J. Chem. *25*, 2535 (1972).
139. *Dickson, R.S., Michel, L.J.:* Aust. J. Chem. *28*, 1943 (1975).
140. *Dickson, R.S., Michel, L.J.:* Aust. J. Chem. *28*, 1957 (1975).
141. *Dobbie, R.C., Mason, P.R.:* J.C.S. Dalton *1973*, 1124.
142. *Dobbie, R.C., Mason, P.R., Porter, R.J.:* Chem. Commun. *1972*, 612.
143. *Dolby, R., Matheson, T.W., Nickolson, B.K., Robinson, B.H., Simpson, J.:* J. Organomet. Chem. *43*, C 13 (1972).
144. *Dolby, R., Robinson, B.H.:* J.C.S. Dalton *1972*, 2046.
145. *Dolby, R., Robinson, B.H.:* J.C.S. Dalton *1973*, 1794.
146. *Domingos, A.J.P., Johnson, B.F.G., Lewis, J.:* J. Organomet. Chem. *36*, C 43 (1972).
147. *Drakesmith, A.J., Whyman, R.:* J.C.S. Dalton *1973*, 362.
148. *Eady, C.R., Jackson, W.G., Johnson, B.F.G., Lewis, J., Matheson, T.W.:* Chem. Commun. *1975*, 958.
149. *Eady, C.R., Johnson, B.F.G., Lewis, J.:* J. Organomet. Chem. *37*, C 39 (1972).
150. *Eady, C.R., Johnson, B.F.G., Lewis, J.:* J. Organomet. Chem. *57*, C 84 (1973).
151. *Eady, C.R., Johnson, B.F.G., Lewis, J.:* J.C.S. Dalton *1975*, 2606.
152. *Eady, C.R., Johnson, B.F.G., Lewis, J., Matheson, T.:* J. Organomet. Chem. *57*, C 82 (1973).
153. *Edwards, D.A.:* Specialist Periodical Reports, Organometallic Chemistry, *3*, 210 (1974), *4*, 199 (1975).
154. *Einstein, F.W.B., Jones, R.D.G.:* J. Chem. Soc. *A 1971*, 3359.
155. *Einstein, F.W.B., Jones, R.D.G.:* Inorg. Chem. 11, 395 (1972).
156. *Einstein, F.W.B., Jones, R.D.G.:* J.C.S. Dalton *1972*, 2563.
157. *Einstein, F.W.B., Pilotti, A.M., Restivo, R.:* Inorg. Chem. *10*, 1947 (1971).
158. *Elder, R.C.:* Inorg. Chem. *13*, 1037 (1974).
159. *Elder, P.A., Robinson, B.H.:* J. Organomet. Chem. *36*, C 45 (1972).
160. *Elder, P.A., Robinson, B.H., Simpson, J.:* J.C.S. Dalton *1975*, 1771.
161. *Evans, G.D., Slater, J., Giusto, D., Sheline, R.K.:* Inorg. Nucl. Chem. Letters 7, 771 (1971).
162. *Evans, J., Johnson, B.F.G., Lewis, J., Matheson, T.W.:* J. Amer. Chem. Soc. *97*, 1245 (1975).
163. *Evans, J., Johnson, B.F.G., Lewis, J., Matheson, T.W.:* J. Organomet. Chem. *97*, C 16 (1975).
164. *Evans, J., Johnson, B.F.G., Lewis, J., Norton, J.R., Cotton, F.A.:* Chem. Commun. *1973*, 807.
165. *Evans, M., Hursthouse, M., Randall, E.W., Rosenberg, E., Milbone, L., Valle, M.:* Chem. Commun. *1972*, 545.
166. *Ferguson, J.A., Meyer, T.J.:* Chem. Commun. *1971*, 623.
167. *Ferguson, J.A., Meyer, T.J.:* J. Amer. Chem. Soc. *94*, 3409 (1972).
168. *Ferrari, R.P., Vaglio, G.A., Gambino, O., Valle, M., Cetini, G.:* J.C.S. Dalton *1972*, 1998.
169. *Ferraris, G., Gervasio, G.:* J.C.S. Dalton *1972*, 1057.
170. *Ferraris, G., Gervasio, G.:* J.C.S. Dalton *1973*, 1933.
171. *Ferraris, G., Gervasio, G.:* J.C.S. Dalton, *1974*, 1813.
172. *Fieldhouse, S.A., Cleland, A.J., Freeland, B.H., Mann, C.D.M., O'Brien, R.J.:* J. Chem. Soc. *A 1971*, 2536.

173. *Fischer, E. O., Däweritz, A.:* Angew. Chem. *87*, 360 (1975), Angew. Chem. Int. Ed. Engl. *14*, 346 (1975).
174. *Fischer, J., Mitschler, A., Weiss, R., Dehand, J., Nennig, J. F.:* J. Organomet. Chem. *91*, C 37 (1975).
175. *Forster, A., Johnson, B. F. G., Lewis, J., Matheson, T. W., Robinson, B. H., Jackson, W. G.:* Chem. Commun. *1974*, 1042.
176. *Frankel, R. B., Reiff, W. M., Bernal, I., Good, M.:* Inorg. Chem. *13*, 493 (1974).
177. *Frankel, R. B., Reiff, W. M., Meyer, T. J., Cramer, J. L.:* Inorg. Chem. *13*, 2515 (1974).
178. *Frisch, P. D., Dahl, L. F.:* J. Amer. Chem. Soc. *94*, 5082 (1972).
179. *Gainsford, G. J., Guss, J. M., Ireland, P. R., Mason, R., Bradford, C. W., Nyholm, R. S.:* J. Organomet. Chem. *40*, C 70 (1972).
180. *Gall, R. S., Chu, C. T. W., Dahl, L. F.:* J. Amer. Chem. Soc. *96*, 4019 (1974).
181. *Gall, R. S., Connelly, N. G., Dahl, L. F.:* J. Amer. Chem. Soc. *96*, 4017 (1974).
182. *Gambino, O., Ferrari, R. P., Chinone, M., Vaglio, G. A.:* Inorg. Chim. Acta *12*, 155 (1975).
183. *Gambino, O., Sappa, E., Cetini, G.:* J. Organomet. Chem. *44*, 185 (1972).
184. *Gambino, O., Vaglio, G. A., Ferrari, R. P., Cetini, G.:* J. Organomet. Chem. *30*, 381 (1971).
185. *Gambino, O., Valle, M., Aime, S., Vaglio, G. A.:* Inorg. Chim. Acta *8*, 71 (1974).
186. *Gansow, O., Burke, A. R., Vernon, W. D.:* J. Amer. Chem. Soc. *94*, 2550 (1972).
187. *Gardner, P. J., Cartner, A., Cunninghame, R. C., Robinson, B. H.:* J.C.S. Dalton *1975*, 2582.
188. *Gardner, S. A., Andrews, P. S., Rausch, M. D.:* Inorg. Chem. *12*, 2396 (1973).
189. *Gardner, S. A., Rausch, M. D.:* J. Organomet. Chem. *78*, 415 (1974).
190. *Gervasio, G., Ferraris, G.:* Cryst. Struct. Comm. *3*, 447 (1973).
191. *Gilmore, C. J., Woodward, P.:* J. Chem. Soc. *A 1971*, 3453.
192. *Grant, S., Newman, J., Manning, A. R.:* J. Organomet. Chem. *96*, C 11 (1975).
193. *Green, M., Howard, J. A., Spencer, J. L., Stone, F. G. A.:* Chem. Commun. *1975*, 3.
194. *Guss, J. M., Mason, R., Søtofte, I., van Koten, G., Noltes, J. G.:* Chem. Commun. *1972*, 446.
195. *Guss, J. M., Mason, R., Thomas, K. M., van Koten, G., Noltes, J. G.:* J. Organomet. Chem. *40*, C 79 (1972).
196. *Haines, R. J., de Beer, J. A., Greatrex, R.:* J. Organomet. Chem. *55*, C 30 (1973).
197. *Haines, R. J., de Beer, J. A., Greatrex, R.:* J. Organomet. Chem. *85*, 89 (1975).
198. *Hall, M. B., Fenske, R. F.:* Inorg. Chem. *11*, 768 und 1619 (1972).
199. *Hallgren, J. E., Eschbach, C. S., Seyferth, D.:* J. Amer. Chem. Soc. *94*, 2547 (1972).
200. *Hamer, A. D., Walton, R. A.:* Inorg. Chem. *13*, 1446 (1974).
201. *Haupt, H. J., Neumann, F.:* J. Organomet. Chem. *74*, 185 (1974).
202. *Haupt, H. J., Neumann, F., Preut, H.:* J. Organomet. Chem. *99*, 439, (1975).
203. *Heaton, B. T., Towl, A. D. C., Chini, P., Fumagalli, A., McCaffrey, D. J. A., Martinengo, S.:* Chem. Commun. *1975*, 523.
204. *Herberhold, M., Süß, G.:* Angew. Chem. *87*, 710 (1975), Angew. Chem. Int. Ed. Engl. *14*, 700 (1975).
205. *Hidai, M., Kokura, M., Uchida, Y.:* J. Organomet. Chem. *52*, 431 (1973).
206. *Hollander, F. J., Coucouvanis, D.:* J. Amer. Chem. Soc. *96*, 5646 (1974).
207. *Holm, R. H.:* Endeavour *34*, 38 (1975).
208. *Hosseini, H. E., Nixon, J. F.:* J. Organomet. Chem. *97*, C 24 (1975).
209. *Howard, J. A. K., Knox, S. A. R., Riera, V., Stone, F. G. A., Woodward, P.:* Chem. Commun. *1974*, 452.
210. *Howard, J. A. K., Knox, S. A. R., Stone, F. G. A., Szary, A. C., Woodward, P.:* Chem. Commun. *1974*, 788.
211. *Howard, J. A. K., Woodward, P.:* J. Chem. Soc. *A 1971*, 3648.
212. *Howell, J. A. S., Matheson, T. W., Mays, M. J.:* J. Organomet. Chem. *88*, 363 (1975).
213. *Hsieh, A. T. T., Knight, J.:* J. Organomet. Chem. *26*, 125 (1971).
214. *Hsieh, A. T. T., Mays, M. J.:* J. Organomet. Chem. *37*, C 53 (1972).
215. *Hsieh, A. T. T., Mays, M. J.:* J. Organomet. Chem. *39*, 157 (1972).
216. *Huie, B. T., Knobler, C. B., Kaesz, H. D.:* Chem. Commun. *1975*, 684.

217. *Humphries, A.P., Knox, S.A.R.:* J.C.S. Dalton *1975*, 1710.
218. *Huttner, G., Lorenz, H.:* Chem. Ber. *107*, 996 (1974).
219. *Huttner, G., Lorenz, H.:* Chem. Ber. *108*, 973 (1975).
220. *Ingle, W.M., Preti, G., MacDiarmid, A.G.:* Chem. Commun. *1973*, 497.
221. *Iwatate, K., Dasgupta, S.R., Schneider, R.L., Smith, C.G., Watters, K.L.:* Inorg. Chim. Acta *15*, 191 (1975).
222. *Jackson, W.G., Johnson, B.F.G., Kelland, J.W., Lewis, J., Schorpp, K.T.:* J. Organomet. Chem. *87*, C 27 (1975).
223. *Jackson, W.G., Johnson, B.F.G., Kelland, J.W., Lewis, J., Schorpp, K.T.:* J. Organomet. Chem. *88*, C 17 (1975).
224. *Jackson, W.G., Johnson, B.F.G., Lewis, J.:* J. Organomet. Chem. *90*, C 13 (1975).
225. *Jarvis, J.A., Kilbourn, B.T., Pearce, R., Lappert, M.F.:* Chem. Commun. *1973*, 475.
226. *Johnson, B.F.G., Lewis, J.:* Acc. Chem. Res. *1*, 245 (1968).
227. *Johnson, B.F.G., Lewis, J.:* Inorg. Synth. *13*, 92 (1972).
228. *Johnson, B.F.G., Lewis, J., Matheson, T.W.:* Chem. Commun. *1974*, 441.
229. *Johnson, B.F.G., Lewis, J., Schorpp, K.T.:* J. Organomet. Chem. *91*, C 13 (1975).
230. *Johnson, B.F.G., Lewis, J., Twigg, M.V.:* J.C.S. Dalton *1975*, 1876.
231. *Johnston, R.D.:* Adv. Inorg. Chem. Radiochem. *13*, 471 (1970).
232. *Jones, C.J., McCleverty, J.A., Orchard, D.G.:* J. Organomet. Chem. *26*, C 19 (1971).
233. *Kaesz, H.D.:* Chem. in Britain *9*, 344 (1973).
234. *Kaesz, H.D., Knox, S.A.R., Koepke, J.W., Saillant, R.B.:* Chem. Commun. *1971*, 477.
235. *Kamijo, T., Kitamura, T., Sakamoto, N., Joh, T.:* J. Organomet. Chem. *54*, 265 (1973).
236. *Kamijyo, N., Watanabe, T.:* Bull. Chem. Soc. Jap. *47*, 373 (1974).
237. *Kanter, H., Dimroth, K.:* Tetrahedron Lett. *1975*, 545.
238. *Karel, K.J., Norton, J.R.:* J. Amer. Chem. Soc. *96*, 6812 (1974).
239. *Keister, J.B., Shapley, J.R.:* J. Organomet. Chem. *85*, C 29 (1975).
240. *Khand, I.U., Knox, G.R., Pauson, P.L., Watts, W.E.:* Chem. Commun. *1971*, 36.
241. *King, R.B.:* Progr. Inorg. Chem. *15*, 287 (1972).
242. *King, R.B., Ackermann, M.N.:* J. Organomet. Chem. *67*, 431 (1974).
243. *King, R.B., Braitsch, D.M., Kapoor, P.N.:* Chem. Commun. *1972*.
244. *King, R.B., Braitsch, D.M., Kapoor, P.N.:* J. Amer. Chem. Soc. *97*, 60 (1975).
245. *King, R.B., Kapoor, P.N.:* Inorg. Chem. *11*, 336 (1972).
246. *Kirtley, S.W., Olsen, J.P., Bau, R.:* J. Amer. Chem. Soc. *95*, (1973).
247. *Kitamura, T., Joh, T.:* J. Organomet. Chem. *65*, 235 (1974).
248. *Knight, J., Mays, M.J.:* Chem. Commun. *1971*, 62.
249. *Knight, J., Mays, M.J.:* J.C.S. Dalton *1972* , 1022.
250. *Knox, S.A.R., Kaesz, H.D.:* J. Amer. Chem. Soc. *93*, 4594 (1971).
251. *Knox, S.A.R., Koepke, J.W., Andrews, M.A., Kaesz, H.D.:* J. Amer. Chem. Soc. *97*, 3942 (1975).
252. *Knox, G.R., Pryde, A.:* J. Organomet. Chem. *74*, 105 (1974).
253. *Knox, S.A.R., Stone, F.G.A.:* J. Chem. Soc. *A 1971*, 2874.
254. *Koepke, J.W., Johnson, J.R., Knox, S.A.R., Kaesz, H.D.:* J. Amer. Chem. Soc. *97*, 3947 (1975).
255. *Koerner von Gustorf, E., Wagner, R.:* Angew. Chem. *83*, 968 (1971); Angew. Chem. Int. Ed. Engl. *10*, 910 (1971).
256. *v. Koten, G., Leusink, A.J., Noltes, J.G.:* Inorg. Nucl. Chem. Letters 7, 227 (1971).
257. *v. Koten, G., Leusink, A.J., Noltes, J.G.:* J. Organomet. Chem. *84*, 117 (1975).
258. *v. Koten, G., Leusink, A.J., Noltes, J.G.:* J. Organomet. Chem. *85*, 105 (1975).
259. *v. Koten, G., Noltes, J.G.:* Chem. Commun. *1972*, 59.
260. *v. Koten, G., Noltes, J.G.:* Chem. Commun. *1972*, 452.
261. *v. Koten, G., Noltes, J.G.:* Chem. Commun. *1974*, 575.
262. *v. Koten, G., Noltes, J.G.:* J. Organomet. Chem. *82*, C 53 (1974).
263. *v. Koten, G., Noltes, J.G.:* J. Organomet. Chem. *84*, 129 (1975).

264. *v. Koten, G., Noltes, J. G.:* J. Organomet. Chem. *84*, 419 (1975).
265. *v. Koten, G., Noltes, J. G.:* J. Organomet. Chem. *102*, 551 (1975).
266. *Kristoff, J. S., Shriver, D. F.:* Inorg. Chem. *13*, 499 (1974).
267. *Kudo, K., Hudai, M., Uchida, Y.:* J. Organomet. Chem. *33*, 393 (1971).
268. *Labrue, D., Poilblanc, R.:* Inorg. Chim. Acta *6*, 387 (1972).
269. *Lappert, M. F., Pearce, R.:* Chem. Commun. *1973*, 24.
270. *Lawton, S. L., Rohrbaugh, W. J., Kokotailo, G. T.:* Inorg. Chem. *11*, 612 (1972).
271. *Lessmeister, P., Schäfer, H.:* Z. anorg. allg. Chem. *417*, 171 (1975).
272. *Leusink, A. J., v. Koten, G., Noltes, J. G.:* J. Organomet. Chem. *56*, 379 (1973).
273. *Lewis, H. C., Storhoff, B. N.:* J. Organomet. Chem. *43*, 1 (1972).
274. *Longoni, G., Chini, P., Lower, L. D., Dahl, L. F.:* J. Amer. Chem. Soc. *97*, 5034 (1975).
275. *Malatesta, L., Caglio, G., Angoletta, M.:* Inorg. Synth. *13*, 95 (1972).
276. *Markó, L., Markó, B.:* Inorg. Chim. Acta *14*, L 39 (1975).
277. *Martinengo, S., Chini, P.:* Gazzetta *102*, 344 (1972).
278. *Martinengo, S., Chini, P., Albano, V. G., Cariati, F., Salvatori, T.:* J. Organomet. Chem. *59*, 379 (1973).
279. *Martinengo, S., Chini, P., Giordano, G.:* J. Organomet. Chem. *27*, 389 (1971).
280. *Martinengo, S., Chini, P., Giordano, G., Ceriotti, A., Albano, V. G., Ciani, G.:* J. Organomet. Chem. *88*, 375 (1975).
281. *Mason, R.:* Pure Appl. Chem. *33*, 513 (1973).
282. *Mason, R., Mingos, D. M. P.:* J. Organomet. Chem. *50*, 53 (1973).
283. *Mason, R., Thomas, K. M.:* J. Organomet. Chem. *43*, C 39 (1972).
284. *Mason, R., Thomas, K. M., Mingos, D. M. P.:* J. Amer. Chem. Soc. *95*, 3802 (1973).
285. *Mason, R., Zubieta, J. A.:* J. Organomet. Chem. *66*, 289 (1974).
286. *Mason, R., Zubieta, J., Hsieh, A. T. T., Knight, J., Mays, M. J.:* Chem. Commun. *1972*, 200.
287. *Matheson, T. W., Peake, B. M., Robinson, B. H., Simpson, J., Watson, D. J.:* Chem. Commun. *1973*, 894.
288. *Matheson, T. W., Robinson, B. H.:* J. Organomet. Chem. *88*, 367 (1975).
289. *Matheson, T. W., Robinson, B. H., Tham, W. S.:* J. Chem. Soc. *A 1971*, 1457.
290. *Mawby, R. J.:* Specialist Periodical Reports, Organometallic Chemistry, *1*, 228 (1972), *2*, 273 (1973).
291. *Mercer, G. D., Shu, J. S., Rauchfuss, T. B., Roundhill, D. M.:* J. Amer. Chem. Soc. *97*, 1967 (1975).
292. *Meyer, T. J.:* Progr. Inorg. Chem. *19*, 1 (1975).
293. *Milone, L., Aime, S., Randall, E. W., Rosenberg, E.:* Chem. Commun. *1975*, 452.
294. *Mingos, D. M. P.:* J.C.S. Dalton *1974*, 133.
295. *Modinos, A., Woodward, P.:* J.C.S. Dalton *1975*, 1534.
296. *Moelwyn-Hughes, J. T., Garner, A. W. B., Gordon, N.:* J. Organomet. Chem. *26*, 373 (1971).
297. *Muetterties, E. L., Wright, C. M.:* Quart. Rev. Chem. Soc. *21*, 109, (1967).
298. *Müller, J., Dorner, H.:* Angew. Chem. *85*, 867 (1973), Angew. Chem. Int. Ed. Engl. *12*, 843 (1973).
299. *Müller, J., Dorner, H., Huttner, G., Lorenz, H.:* Angew. Chem. *85*, 1117 (1973), Angew. Chem. Int. Ed. Engl. *12*, 1005 (1973).
300. *Müller, J., Dorner, H., Köhler, F. H.:* Chem. Ber. *106*, 1122 (1973).
301. *Müller, J., Schmitt, S.:* J. Organomet. Chem. *97*, C 54 (1975).
302. *Natile, G., Pignataro, S., Innorta, G., Bor, G.:* J. Organomet. Chem. *40*, 215 (1972).
303. *Neuman, M. A., Trinh-Toan, Dahl, L. F.:* J. Amer. Chem. Soc. *94*, 3383 (1972).
304. *Newman, J., Manning, A. R.:* J.C.S. Dalton *1974*, 2549.
305. *Nicholson, B. K., Robinson, B. H., Simpson, J.:* J. Organomet. Chem. *66*, C 3 (1974).
306. *Noltes, J. G.:* J. Organomet. Chem. *100*, 177 (1975).
307. *Norton, J. R., Collman, J. P.:* Inorg. Chem. *12*, 476 (1973).
308. *Norton, J. R., Collman, J. P., Dolcetti, G., Robinson, W. T.:* Inorg. Chem. *11*, 382 (1972).
309. *Otsuka, S., Nakamura, A., Tatsuno, Y., Miki, M.:* J. Amer. Chem. Soc. *94*, 3761 (1972).

310. *Otsuka, S., Tatsuno, Y., Miki, M., Aoki, T., Matsumoto, M., Yoshioka, H., Nakatsu, K.:* Chem. Commun. *1973*, 445.
311. *Paik, H.N., Carty, A.J., Mathew, M., Palenik, G.J.:* Chem. Commun. *1974*, 946.
312. *Palyi, G., Baumgartner, F., Czajlik, I.:* J. Organomet. Chem. *49*, C 85 (1973).
313. *Palyi, G., Piacenti, F., Marko, L.:* Inorg. Chim. Acta Rev. *4*, 109 (1970).
314. *Palyi, G., Vàradi, G.:* J. Organomet. Chem. *86*, 119 (1975).
315. *Peake, B.M., Robinson, B.H., Simpson, J., Watson, D.G.:* Chem. Commun. *1974*, 945.
316. *Penfold, B.R.:* Perspectives in Structural Chemistry *2*, 71 (1968).
317. *Penfold, B.R., Robinson, B.H.:* Acc. Chem. Res. *6*, 73 (1973).
318. *Piacenti, F., Bianchi, M., Frediani, P., Benedetti, E.:* Inorg. Chem. *10*, 2759 (1971).
319. *Poë, A.J., Twigg, M.V.:* J. Organomet. Chem. *50*, C 39 (1973).
320. *Poë, A.J., Twigg, M.V.:* J.C.S. Dalton *1974*, 1860.
321. *Poë, A.J., Twigg, M.V.:* Inorg. Chem. *13*, 2982 (1974).
322. *Pregaglia, G.F., Andreetta, A., Ferrari, G.F., Montrasi, G., Ugo, R.:* J. Organomet. Chem. *33*, 73 (1971).
323. *Preut, H., Haupt, H.J.:* Chem. Ber. *107*, 2860 (1974).
324. *Preut, H., Haupt, H.J.:* Chem. Ber. *108*, 1447 (1975).
325. *Raper, G., McDonald, W.S.:* J. Chem. Soc. *A 1971*, 3430.
326. *Rausch, M.D., Bernal, I., Davis, B.R., Siegel, A., Higbie, F.A., Westover, G.F.:* J. Coord. Chem. *3*, 149 (1974).
327. *Roberts, P.J., Trotter, J.:* J. Chem. Soc. *A 1971*, 1479.
328. *Robinson, B.H., Spencer, J.L.:* J. Chem. Soc. *A 1971*, 2045.
329. *Robinson, B.H., Spencer, J.L.:* J. Organomet. Chem. *30*, 267 (1971).
330. *Robinson, B.H., Spencer, J.L.:* J. Organomet. Chem. *33*, 97 (1971).
331. *Rossetti, R., Stanghellini, P.L., Gambino, O., Cetini, G.:* Inorg. Chim. Acta *6*, 205 (1972).
332. *Rouschias, G.:* Chem. Rev. *74*, 531 (1974).
333. *Ruff, J.K., White, R.P., Dahl, L.F.:* J. Amer. Chem. Soc. *93*, 2159 (1971).
334. *Ryan, R.C., Dahl, L.F.:* J. Amer. Chem. Soc. *97*, 6904 (1975).
335. *Sappa, E., Gambino, O., Cetini, S.:* J. Organomet. Chem. *35*, 375 (1972).
336. *Sappa, E., Gambino, O., Milone, L., Cetini, G.:* J. Organomet. Chem. *39*, 169 (1972).
337. *Sappa, E., Milone, L.:* J. Organomet. Chem. *61*, 383 (1973).
338. *Schäfer, H., von Schnering, H.G.:* Angew. Chem. *76*, 833 (1964).
339. *Schmid, G., Bätzel, V.:* J. Organomet. Chem. *46*, 149 (1972).
339a. *Schmid, G., Bätzel, V., Etzrodt, G., Pfeil, R.:* J. Organomet. Chem. *86*, 257 (1975).
340. *Schmid, G., Stutte, B.:* J. Organomet. Chem. *37*, 375 (1972).
341. *Schwarzhans, K.E., Steiger, H.:* Angew. Chem. *84*, 587 (1972)., Angew. Chem. Int. Ed. Engl. *11*, 535 (1972).
342. *Seitz, L.M., Madl, R.:* J. Organomet. Chem. *34*, 415 (1972).
343. *Seyferth, D., Eschbach, C.S., Nestle, M.O.:* J. Organomet. Chem. *97*, C 11 (1975).
344. *Seyferth, D., Eschbach, C.S., Williams, G.H., Hung, P.L.K., Cheng, Y.M.:* J. Organomet. Chem. *78*, C 13 (1974).
345. *Seyferth, D., Hallgren, J.E.:* J. Organomet. Chem. *49*, C 41 (1973).
346. *Seyferth, D., Hallgren, J.E., Eschbach, C.S.:* J. Amer. Chem. Soc. *96*, 1730 (1974).
347. *Seyferth, D., Hallgren, J.E., Hung, P.L.K.:* J. Organomet. Chem. *50*, 265 (1973).
348. *Seyferth, D., Hallgren, J.E., Spohn, R.J., Williams, G.H., Nestle, M.O., Hung, P.L.K.:* J. Organomet. Chem. *65*, 99 (1974).
349. *Seyferth, D., Hung, P.L.K., Hallgren, J.E.:* J. Organomet. Chem. *44*, C 55 (1972).
350. *Seyferth, D., Millar, M.D.:* J. Organomet. Chem. *38*, 373 (1972).
351. *Seyferth, D., Spohn, R.J., Hallgren, J.E.:* J. Organomet. Chem. *28*, C 34 (1971).
352. *Seyferth, D., Williams, G.H.:* J. Organomet. Chem. *38*, C 11 (1972).
353. *Seyferth, D., Williams, G.H., Hallgren, J.E.:* J. Amer. Chem. Soc. *95*, 266 (1973).
354. *Seyferth, D., Williams, G.H., Hung, P.L.K., Hallgren, J.E.:* J. Organomet. Chem. *71*, 97 (1974).

355. *Seyferth, D., Williams, G.H., Traficante, D.D.:* J. Amer. Chem. Soc. *96*, 604 (1974).
356. *Seyferth, D., Williams, G.H., Wehman, A.T., Nestle, M.O.:* J. Amer. Chem. Soc. *97*, 2107 (1975).
357. *Shapley, J.R., Keister, J.B., Churchill, M.R., DeBoer, B.G.:* J. Amer. Chem. Soc. *97*, 4145 (1975).
358. *Shapley, J.R., Richter, S.I., Tachikawa, M., Keister, J.B.:* J. Organomet. Chem. *94*, C 43 (1975).
359. *Sheldrick, G.M., Yesinowski, J.P.:* J.C.S. Dalton *1975*, 873.
360. *Shriver, D.F., Lehman, D., Strope, D.:* J. Amer. Chem. Soc. *97*, 1594 (1975).
361. *Simon, G.L., Dahl, L.F.:* J. Amer. Chem. Soc. *95*, 2164 (1973).
362. *Simon, G.L., Dahl, L.F.:* J. Amer. Chem. Soc. *95*, 2175 (1973).
363. *Sisak, A., Sisak, C., Ungvary, F., Palyi, G., Marko, L.:* J. Organomet. Chem. *90*, 77 (1975).
364. *Smith, G.C., Chojnacki, T.P., Dasgupta, S.R., Iwatate, K., Watters, K.L.:* Inorg. Chem. *14*, 1419 (1975).
365. *Spiro, T.G.:* Progr. Inorg. Chem. *11*, 1 (1970).
366. *Stevenson, D.L., Wei, C.H., Dahl, L.F.:* J. Amer. Chem. Soc. *93*, 6027 (1971).
367. *Stewart, R.P., Anders, U., Graham, W.A.G.:* J. Organomet. Chem. *32*, C 49 (1971).
368. *Strouse, C.E., Dahl, L.F.:* J. Amer. Chem. Soc. *93*, 6032 (1971).
369. *Stuntz, G.F., Shapley, J.R.:* Inorg. Nucl. Chem. Letters *12*, 49 (1976).
370. *Symon, D.A., Waddington, T.C.:* J.C.S. Dalton *1975*, 2140.
371. *Tachikawa, M., Shapley, J.R., Pierpont, C.G.:* J. Amer. Chem. Soc. *97*, 7172 (1975).
372. *Taylor, N.J., Chieh, P.C., Carty, A.J.:* Chem. Commun. *1975*, 448.
373. *Teo, B.K., Hall, M.B., Fenske, R.F., Dahl, L.F.:* J. Organomet. Chem. *70*, 413 (1974).
374. *Treichel, P.M., Dean, W.K., Douglas, W.M.:* Inorg. Chem. *11*, 1609 (1972).
375. *Trinh-Toan, Fehlhammer, W.P., Dahl, L.F.:* J. Amer. Chem. Soc. *94*, 3389 (1972).
376. *Tripathi, S.C., Srivastava, S.C., Mani, R.P., Shrimal, A.K.:* Inorg. Chim. Acta *15*, 249 (1975).
377. *Udovich, C.A., Clark, R.J.:* J. Organomet. Chem. *36*, 355 (1972).
378. *Ungvary, F., Marko, L.:* J. Organomet. Chem. *71*, 283 (1974).
379. *Vaglio, G.A., Gambino, O., Ferrari, R.P., Cetini, G.:* Inorg. Chim. Acta 7, 193 (1972).
380. *Vahrenkamp, H.:* Chem. in uns. Zeit *8*, 112 (1974).
381. *Vahrenkamp, H.:* Angew. Chem. *87*, 363 (1975), Angew. Chem. Int. Ed. Engl. *14*, 322 (1975).
382. *Valle, M., Gambino, O., Milone, L., Vaglio, G.A., Cetini, G.:* J. Organomet. Chem. *38*, C 46 (1972).
383. *Vergamini, P.J., Vahrenkamp, H., Dahl, L.F.:* J. Amer. Chem. Soc. *93*, 6326 (1971).
384. *Vergamini, P.J., Vahrenkamp, H., Dahl, L.F.:* J. Amer. Chem. Soc. *93*, 6327 (1971).
385. *Vizi-Orosz, A., Palyi, G., Marko, L.:* J. Organomet. Chem. *60*, C 25 (1973).
386. *Volhardt, K.P.C., Bercaw, J.E., Bergman, R.G.:* J. Amer. Chem. Soc. *96*, 4998 (1974).
387. *Volhardt, K.P.C., Bercaw, J.E., Bergman, R.G.:* J. Organomet. Chem. *97*, 283 (1975).
388. *Voyevodskaya, T.I., Pribytkova, I.M., Ustiynyuk, Y.A.:* J. Organomet. Chem. *37*, 187 (1972).
389. *Wade, K.:* Chem. Commun. *1971*, 792.
390. *Wade, K.:* Inorg. Nucl. Chem. Letters *8*, 559 (1972).
391. *Wade, K.:* Inorg. Nucl. Chem. Letters *8*, 563 (1972).
392. *Wade, K.:* Inorg. Nucl. Chem. Letters *8*, 823 (1972).
393. *Wade, K.:* Chem. in Britain *11*, 177 (1975).
394. *White, J.W., Wright, C.J.:* J. Chem. Soc. *A 1971*, 2843.
395. *Whyman, R.:* J. Organomet. Chem. *29*, C 36 (1971).
396. *Whyman, R.:* J.C.S. Dalton *1972*, 1375.
397. *Whyman, R.:* J.C.S. Dalton *1972*, 2294.
398. *Whyman, R.:* J. Organomet. Chem. *56*, 339 (1973).
399. *Whyman, R.:* J. Organomet. Chem. *63*, 467 (1973).

400. *Yamamoto, T., Garber, A.R., Bodner, G.M., Todd, L.J.:* J. Organomet. Chem. *56*, C 23 (1973).
401. *Yang, C.Y., Johnson, K.H., Holm, R.H., Norman, J.G.:* J. Amer. Chem. Soc. *97*, 6596 (1975).
402. *Yasufuku, K., Aoki, K., Yamazaki, H.:* Bull. Chem. Soc. Jap. *48*, 1616 (1975).
403. *Yasufuku, K., Yamazaki, H.:* Bull. Chem. Soc. Jap. *45*, 2664 (1972).
404. *Yawney, D.B.W., Doedens, R.J.:* Inorg. Chem. *11*, 838 (1972).
405. *Yesinowski, J.P., Bailey, D.:* J. Organomet. Chem. *65*, C 27 (1974).
406. *Yin, C.C., Deeming, A.J.:* J.C.S. Dalton *1974*, 1013.
407. *Yin, C.C., Deeming, A.J.:* J.C.S. Dalton *1975*, 2091.
408. *Young, D.A.:* Inorg. Chem. *12*, 482 (1973).

A Generalized Configuration-Dependent Band Model for Lanthanide Compounds and Conditions for Interconfiguration Fluctuations

J.A. Wilson

Bell Laboratories, Murray Hill, New Jersey 07974.

Table of Contents

1. The Construction and Interpretation of the Model

1.1 On Forming a One Electron Model – Mainly Concerning the d-Band

Well over fifty papers now exist concerning the interconfiguration fluctuation (ICF) rare earth compounds[1]), of which SmB_6 archetype, and there is no lack of models, complex or primitive, attempting to interpret the phenomenon, in all its various magnetic, electrical and structural aspects (*1*). However, there does seem to be lacking a wide-ranging yet detailed model designed to demonstrate why it is the various electronic states (namely, fully "expanded", ICF, or fully "collapsed") are found in the particular compounds in which they are.

Qualitatively, the most transparent type of model, as ever, would be a one-electron model that is capable of rendering both the ground state and, to a high degree, its excitation properties. However, in the present case, accommodations are called for, on both aspects, that are not trivial. These we will try to pursue and represent within the present one-electron-type framework as closely as possible. In seeking to develop the present model, we base it as firmly as possible on the available data, optical, photoemission, electrical, structural, etc. Much of this data is still open to interpretation, and many of the interpretations to follow are made in the light of experience gained with transition metal compounds (*2*).

A generalized approach to the full range of binary RE compounds is not new. Hulliger in his survey of 1968 (*3*) pointed out a great many compounds, such as CeP, $NdBr_2$, TbN, TmTe, YbH_2, that could well develop configurational "ambiguities": many of those suggestions remain still unpursued. Here it is intended largely to restrict our attention to the monochalcogenides, as this is where the mass of new data lies, and also where the likelihood of establishing a satisfactory band model is greatest, the crystal structure being the simplest and the energetics the most favorable. The model generalizes quite readily to the other families of RE compounds. It is presented for the monosulphides in Fig. 1 a–c.

Much in evidence in Fig. 1 a is the characteristic double-seated 'shape' of *f*-shell ionization and excitation, as dictated by Hund's rules, by an on-site exchange stabilization of up to 3 – 1/2 eV, and by an effective absence of crystal field effects upon *f*-state degeneracy. Excitation energies of the type $f^n \rightarrow f^{n-1} d^1$ have been analyzed at length by *Jorgensen* (*4*). Table 6.1 in Ref. (*4b*) gives such energies both for RE^{2+} and for RE^{3+} in CaF_2 [2]). The observed large increase in energy for the trivalent series over the divalent (~ 6 eV) is a central feature of RE chemistry. This is due, – as likewise is the virtual limitation of RE valencies to not more than 3, – to the spatially inner-core unshielded nature of RE *f*-electron wavefunctions. The associated decre-

[1]) We have taken care throughout to avoid the misleading term "mixed valence" which is commonly in use for materials in this state.

[2]) These are plotted in Appendix A, Fig. 6.

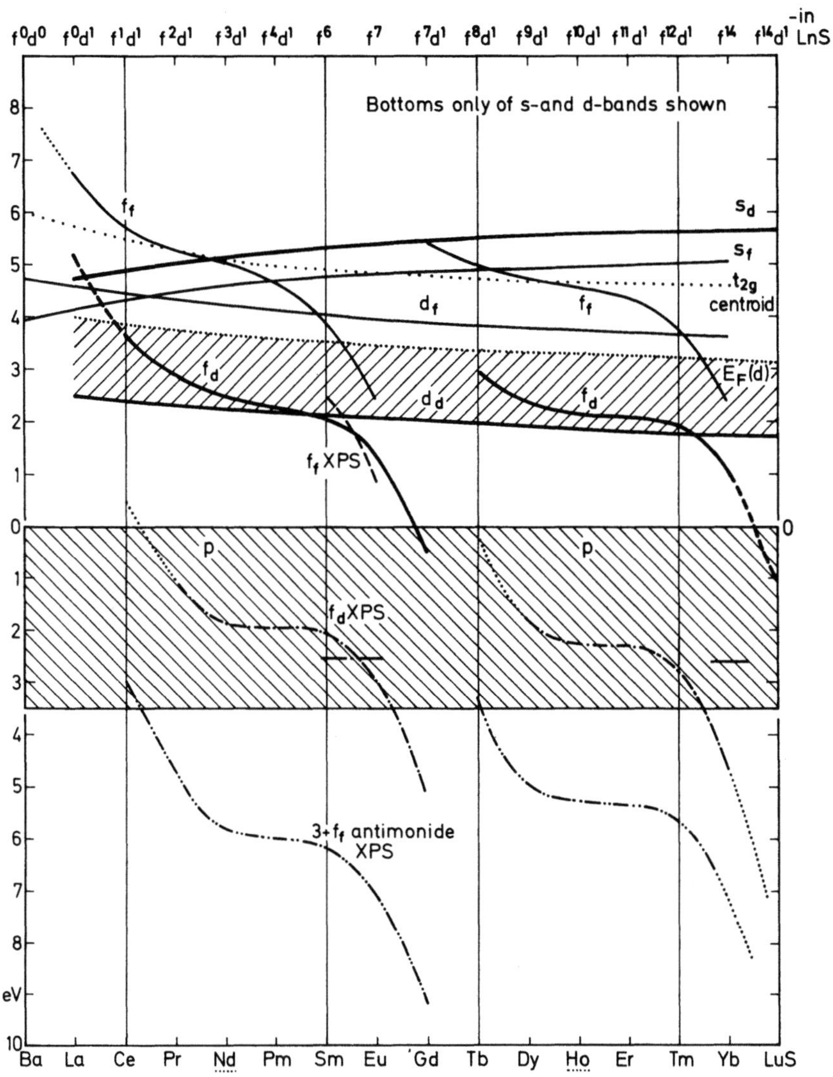

Fig. 1a. The positions of certain one-electron band features, as deduced from optical, XPS, electrical data, etc., are shown across the R.E. monosulphides. The energy positions relating to the alternative configurations $d^1 f^{n-1}$ and $d^0 f^n$ are significantly different, and are shown superposed, in heavy and light (full) lines respectively. Subscripts d and f mark those conditions respectively. The top of the p-band has been selected as common reference level. The d-bands are represented only up to EF_d and the t_{2g_f} centroid respectively (shown dotted). There is a marked drop in the d-band on moving to the collapsed configuration, together with a shift down and to the right in the f-band. Also included are the observed f-electron photoemission results (for a ground-state residue), given by - - - - - for f^n sulphides, — · — · — for $f^{n-1} d^1$ sulphides, and — ·· — ·· for f^{n-1} antimonides. Data was extracted from many references, especially (*9*), (*29*) and (*31*), (see Appendix A 3).

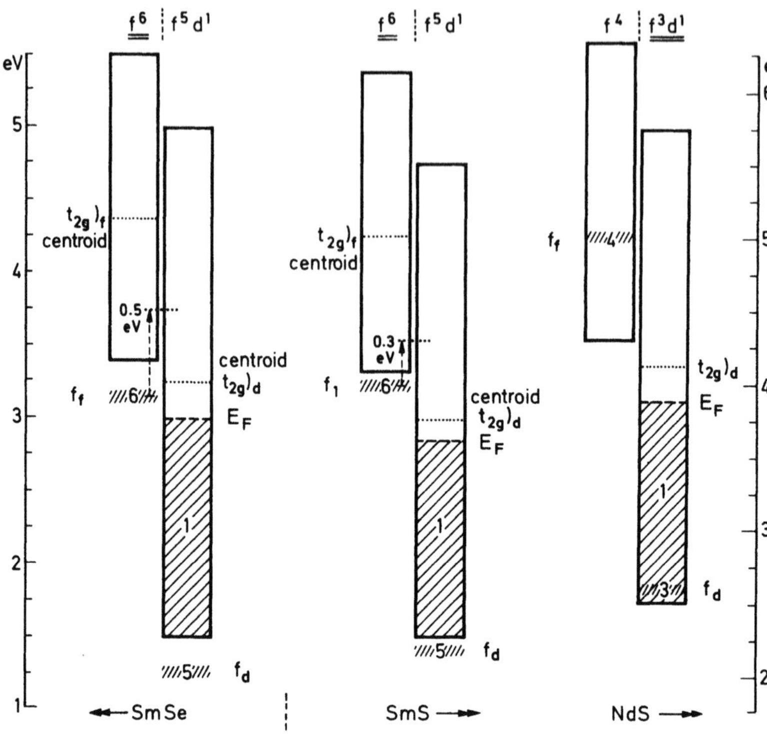

Fig. 1b. Detailed position of t_{2g} and f-bands in the alternative configurations f^n and $f^{n-1}d^1$ for SmS, SmSe and NdS. The $f \rightarrow t_{2g}$ transition is taken as occurring to a t_{2g} centroid appropriate to 50% of bulk collapse. 0.5 eV from the exciting photon is on average lost to the residual f^{n-1} set in the case of Sm, leading to the $f \rightarrow d$ excitation values shown. Other symbols have meaning given in Fig. 1. (N.B. The SmSe scale is shifted by 3/4 eV.)

mental reduction in cationic radius with valence, whether from 2+ to 3+ or 3+ to 4+, is much greater than is found normally for a transition metal, (-special crystal field and high-spin/low-spin effects excluded), e.g. EuF_2 to EuF_3, 21%; CeF_3 to CeF_4, 12%.

Fig. 1, it is to be noted directly, is formed from a superposition of two sets of levels, one set for the 'expanded' f^n configuration, the other for the fully 'collapsed' $f^{n-1}d^1$ configuration. In either condition the materials exhibit a fair amount of p/d and p/s covalency. In the RE monosulphides, a complete collapse would entail a diminution ~ 6% in a_0 [3]). The discontinuous partial collapse into the metallic ICF regime in the much publicized case of SmS doped with Gd (Fig. 2a–c), (which occurs at ~ 16% Gd at 300 K), constitutes in fact only about 50% of this (*5*). The two super-

[3]) See Fig. 1 of Ref. (*33*). Note Fig. 4 of Ref. (*1*) is out of date.

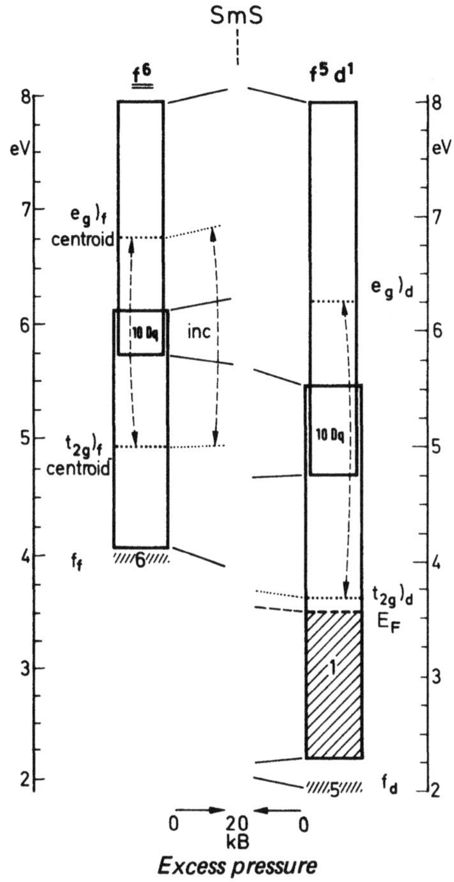

Fig. 1c. Postulated movements of the various f, t_{2g} and e_g band features under pressure, for the two terminal configurations f^6 and $f^5 d^1$ in SmS. The latter condition holds for P $\gtrsim$ 50 kB.

posed halves to Fig. 1a have been united through the ascription of a common top to their p-bands. Since this part of the p-band is derived from non-bonding sulphur orbitals, it should be the least affected part of that band, consequent upon change from the one configuration, the one lattice parameter, to the other. We are hereby cancelling the changes in Madelung Potential that accrue from the opposing elements of decreased a_0 and increased covalency.

Of the two s-bands now to arise, that for the $f^{n-1} d^1$ configuration appears uppermost because of greater $s - p$ interaction. Both s-bands climb through their respective RE series, following the steady "lanthanide contraction" in a_0; Δa_0 from BaS to YbS, and from LaS to LuS, is of 10% and 9% respectively. Energetically as structurally, EuS is more like SrS, YbS more like CaS, than either is like BaS.

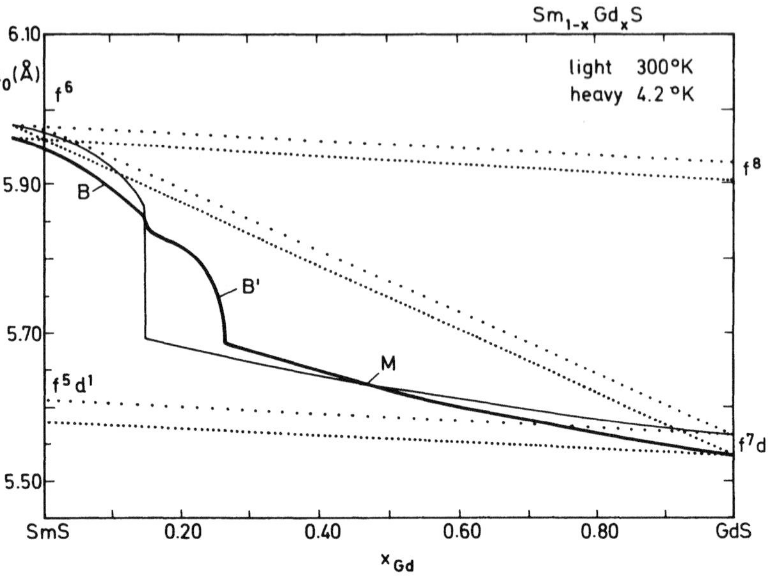

Fig. 2a. Lattice parameters for $Sm_{1-x}Gd_xS$ system at 298 K and 4 K (from Ref. *5*), and indicating lines appropriate to a Vegard's Law analysis of the average configuration. For this particular system (Sm/Gd) S, the lattice parameter after collapse is almost as in pure SmS after collapse under pressure, viz. 5.66 Å at 7 kB (*5b*).

The *d*-bands are much more differentiated (than are the *s*) between the two configurations (1 – 1/2 eV in t_{2g} vs 1/2 eV in *s*), because of the spatially more sensitive tight-binding nature to *d*-function overlap and bonding. In Fig. 1a only the bottoms of the two 't_{2g}' bands have been shown, (going up to the Fermi level in the $f^{n-1}\,d^1$ case, and to the band centroid in the f^n case). Being t_{2g}-based and non-bonding, these t_{2g} features fall only slightly (~ 3/4 eV) across the lanthanide series as the overall *d*-band widens. By contrast, the upper e_g component of the *d*-band, being antibonding, will, one expects, rise quite appreciably with the steady lattice contraction, for both configurations. Clearly changes in the quantity '10 Dq' in such chalcogenides are largely a measure of *p*/*d* covalency, and not simple electrostatics (*6*). In these compounds we are anticipating that the overall *d*-band width reaches ~ 4 to 6 eV, aided in large measure by *p*/*d* mixing. Nevertheless the width is crucially dependent on a_0, and still not too far at its lower limit from yielding to the Mott insulating condition, (c.f. calculations for VO vs CoO) (*6*). The precise roles of correlation, polaronic coupling, and disorder localization, as they are played in for example, the low conductivity of a d^1-doped *expanded* system like $Sm_{0.95}Nd_{0.05}Se$ (*7*) are yet to be clarified. That particular topic will be addressed later. In the fully collapsed phase, the occupied width within the lower 't_{2g}' bands is probably ~ 1 – 1/2 eV, much as in $5d^1$(1 T)–TaS_2 (*8a*), or ReO_3 (*8b*).

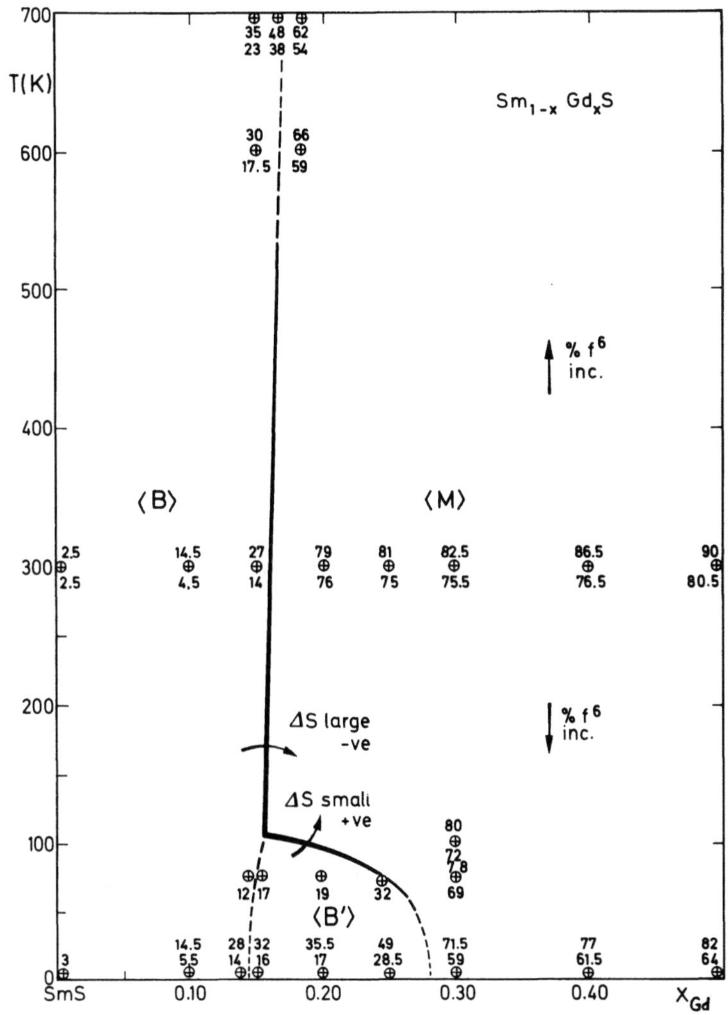

Fig. 2b. Results of Vegard's Law analysis displayed on phase diagram of Ref. 5; upper number is percentage of d^1 sites (Gd and collapsed Sm) in mixture; lower number is percentage of Sm sites in mixture that stand converted to $f^5 d^1$ condition. Note, latter percentages are rather independent of temperature in B phase, but this is not so in M phase, changes there being especially rapid below 100 K. Above about 400 K, the M and B phases slowly converge, as in particular the M phase loses much $f^{n-1} d^1$ representation; $T_{critical} \approx 700$ K.

For any particular lanthanide, upon subjection to *full* collapse, the marked drop (~ 2 eV) that is incurred in the position of the bottom of the t_{2g} band is coupled intimately with the strong contraction in a_0. Firstly, the *d*-function, given its new-found occupancy, becomes more tightly bound, just as for d^1 VO_2 vs d^0 TiO_2,

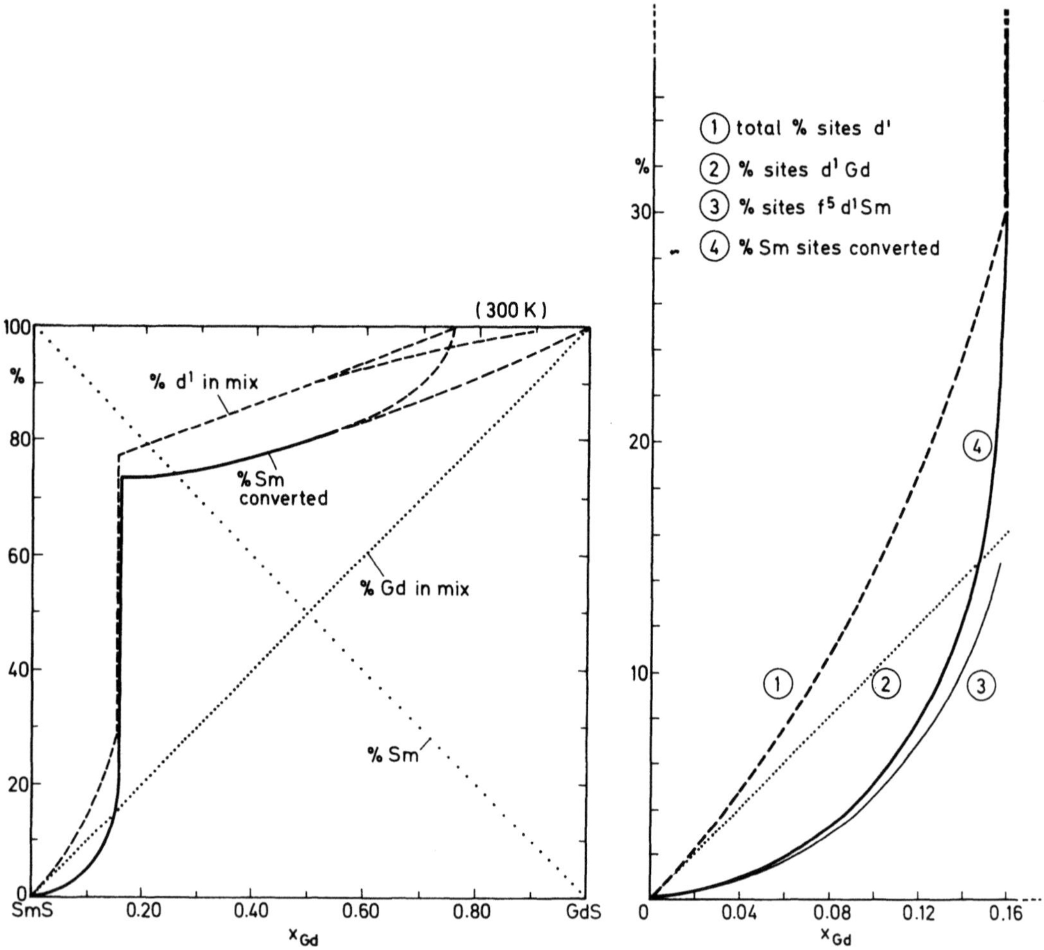

Fig. 2 c. (*1, 2*) Above percentages for 300 K in graph form, showing slow take-off in fraction of Sm standing converted, and its rapid run-away just prior to first-order transition. (For $x > 0.50$ there is uncertainty on the exact course of these lines due to lack of precision here in Fig. 2a.)

or d^1 Ti_2O_3 vs d^0 Sc_2O_3 (*2*). Furthermore, in the present case, this is augmented by the diminished screening afforded the *d*-state by the spatially semi-core *f*-set, now depleted by one. In spite of this, tightening, note however that the increased *p*/*d* interaction[4]), which arises in the collapsed condition, serves more than to sustain the width of the '*d*'-band, and the wavefunctions of this band remain quite highly dispersed, spatially as energetically.

[4]) See Ref. (*10*) on 5 d^1 ReO_3 for a ℓ-character decomposition of V.B and C.B.

1.2 Concerning the Interpretation and Representation of $f \rightarrow d$ Transitions

Through optical work on the monochalcogenides of Eu (*9a*), Yb (*9b*) and Sm (*9c,d*) it is well established that these particular f^n materials have $f \rightarrow d$ *excitation* energies which increase in the order –O, –S, –Se, –Te. This very unusual ordering customarily has been ascribed to an upward shift in the t_{2g} state, originating from a reduction in '10 Dq' with the increase in a_0. Such a description is however surely inappropriate for $f \rightarrow d$ excitations in the present chalcogenides.

$f \rightarrow d$ transitions in RE f^n salts, the above chalcogenides included, have a strong intracation excitonic nature. (The absorption peaks often show, for example, the many J components of the strongly coupled set of $(n - 1)$ *f*-electrons as formed residual to the transition.) In (probable) contrast with the $p \rightarrow d$ transitions, the dispersion structure of the *d*-bands is not made evident in the $f \rightarrow d$ transitions of the RE monochalcogenides, and indeed in basic character, if not in absolute energies, the spectrum is much the same say in EuS as in SrS : Eu (*9a*).

Such $f \rightarrow d$ transitions normally show the characteristic behaviour of ligand-field type transitions (more familiar as $d \rightarrow d$). In particular, with decreasing electronegativity of the associated ligand, there occurs a show 'nephelauxetic' reduction in term separations (*4b*)[5]. This decrease progresses in normal fashion through the halides, but then with the chalcogenides, one encounters the novel reversal noted above. It is clear, from the continued reduction observed in isomer shift through the chalcogenide series (*35*), that a general nephelauxetic effect does still proceed. However, the *differential* effect between the *f*- and *d*-shells has changed sign. Through the chalcogenides, as opposed to the halides, there arises marked p/d hybridization, as the $p - d$ gap diminishes sharply; the dielectric constant rises as a result. The materials are becoming more covalent, with a diminishing Madelung potential. By the stage at which the chalcogenides are reached, the nephelauxetic effect in the *d*-shell is beginning to accelerate away. This retards further expansion of the inner *f*-shell, as the electrons there come now to see proportionately more of the nuclear charge. We note that similar, though less marked, differential nephelauxetic effects have been detected by neutron diffraction and also the Mossbauer effect, between *d*-electrons in t_{2g} and e_g wave functions, (see Ref. (*2a*), § 11). A further marked effect also to derive from the very different radial distributions of the $4f$ and $5d$ wavefunctions will be encountered when discussing relaxation in photoelectron spectra (particularly for the $f^{n-1}d^1$ metals) in § 1.7.

In the case of the f^6 samarium monochalcogenides (*9c,d*), one finds then for the energy of the first $f \rightarrow d$ optical absorption peak – which is associated with the transitions $4f^6(^7F_0) \rightarrow 4f^5(^6H_J)\, t_{2g}^1$ – the following *rising* sequence of values –

SmS 0.8 eV, SmSe 1.0 eV, SmTe 1.2 eV.

[5]) See Appendix A 2.

It is to be emphasized that such transitions involve the full set of coupled on-site f-electrons. The residual f^5 set is left above with the whole series of excited J levels, 5/2 to 15/2, spread over ~ 3/4 eV. The lowest J level probably lies ~ 1/2 eV below the above quoted averaged peak values. The present task is to make some representation of these $f \rightarrow d$ optical transitions on a one-electron basis. This requires consideration of the response in on-site potential under transference of one electron into a t_{2g} wavefunction.

In Fig. 1 we have positioned the $d(t_{2g})$ band in both collapsed and expanded phases (labelled d_a and d_f respectively) at energies that are in line with the "simpler" $p \rightarrow d$ optical "charge transfer" transitions. We assume in these chalcogenides (as in a material like HfS_2, Ref. *8a*) that the d-bands are sufficiently wide for such $p \rightarrow d$ transitions to reflect the d-band dispersion and symmetry characteristics without a large system response around the locale of the excitation. Charge transfer $p \rightarrow d$ energies, unlike the $f \rightarrow d$ transitions introduced above, always decrease in the sequence –S, –Se, –Te. Some $p \rightarrow d$ band *edge* assignments are made in Appendix 3. The $p \rightarrow d_{t_{2g}}$ gap in rocksalts is indirect (Γ to X, Ref. *6*). In Figs. 1b,c we detail examples of the position and width of the d-bands drawn up in line with the present data and overall situation.

$p \rightarrow d$ charge transfer transitions in materials with narrower d-bands begin to lose evidence of d-band dispersion. However the observed transitions seem to occur with energies not differing by too much (~ 0.5 eV) from the calculated one-electron molecular orbital values (*2*). The situation is much as if the d-band had collapsed to its band centroid.

In the case of the $f \rightarrow d$ optical transitions with which we are now most concerned, their essentially intracationic nature ensures that a local treatment of both initial and final states will be required, when attempting to relate the observed transition energies to a one-electron scheme, viz. here from f ground state to t_{2g} band centroid, (remembering the residual f^5-set may be left in an excited condition). Moreover we must note, with this type of transition, that the d-wavefunctions become significantly more tightly bound with the on-site configuration change $f^6\,d^0 \rightarrow f^5\,d^1$, so that we are brought to use, as the appropriate terminal condition, a t_{2g} centroid midway between that for the fully expanded and that for the fully collapsed states. (In the latter state *all* sites of course are in the $f^5\,d^1$ condition, and the lattice itself is contracted.) Following these arguments (and given the 1/2 eV correction appropriate to minimum J in the residual f-set) an observed $f \rightarrow d$ peak energy in SmS of 0.8 eV places the *initial* ground state one-electron f level at a point, immediately below the bottom of the $(t_{2g})_f$ band, as illustrated in Fig. 1b. For SmSe and SmTe, their larger $f \rightarrow d$ transition energies lead to a widening of the gap between this 'f-band' and the bottom of the $(t_{2g})_f$ band.

In SmS electrical data exists which possibly could be taken to enumerate the $f \rightarrow d_f$ gap there as somewhat less than 0.1 eV (*1*). However small instrinsic gaps are notoriously difficult to evaluate electrically. Similarly single crystal I.R. transmission work yields simply a long featureless tail, extending from the 0.8 eV peak towards low energies (Ref. *9e*, Fig. 2), and terminated only below 0.1 eV by what is probably a free-carrier rise coming from some contamination with the $f^5\,d^1$ phase.

The situation for a spontaneous configurational transition will arise if and when the $f \rightarrow (t_{2g})_f$ gap can be brought to zero, though the course of that transition will depend strongly on the precise response of the system, as we shall see in more detail shortly. One way that leads to "collapse" in these materials, it is well known, is the application of pressure. Only 6 – 1/2 kB suffices to bring SmS to its first-order transition. An attempt has been made in Fig. 1c to show the effect of applying pressure upon the various band features. The f^6 set of electrons should be destabilized quite rapidly, as they serve to define the Sm^{2+} radius. Conversely, even though the t_{2g} centroid will rise too, the *bottom* of the t_{2g} band is likely to fall slightly, as the band widens. The larger pressures required to initiate collapse in SmSe and SmTe of up to 40 kB follow from the larger initial gaps to be closed. Optical work under pressure on SmSe and SmTe (*9d*) clearly reveals[6]) that it is not the $f \rightarrow d$ optical transition transition energy that has to be reduced to zero for collapse to start, but the $f \rightarrow (t_{2g})_f$ gap, identified as in Fig. 1b. We shall examine later the effects of correlation and doping on the bottom of the *d*-band, and hence on the progress of the configurational transition.

1.3 Phase Stability and the Number of f-Electrons

Turning from a consideration of relative phase stability as it is anion-determined, let us now examine the much stronger cation dependency. Why is it NdS is collapsed under room conditions, but SmS expanded? The answer lies clearly in the position of the f_f band, as it is found in Fig. 1a (or 1b).

The overall double-seated aspect to its positioning across the lanthanides is clearly a characteristic feature, and one that requires some attention. As we expand on at length in Appendix A1, (following Ref. *4*), the curves are appropriate to loss of an *f*-electron, either completely by photoemission, or, as above, in $f \rightarrow d$ transitions. They will not (without modification) cover $p \rightarrow f$ charge transfer transitions, but these are normally very weak (being forbidden), and fall at energies such as to be overlayed by stronger transitions, including in the f^n materials, the $f \rightarrow d$ transitions[7]). The detailed shape of the *f*-band as constructed is in essence an expression of Hund's rules in conjunction with the steadily increasing binding of the $4f$ electrons across the lanthanides.

La^{2+} and Gd^{2+} are seen to be particularly unstable in the f^n configuration, and this is true even as free ions. In GdS the configuration $f^7\,d^1$ is strongly favored over

---

[6]) The measured pressure shift ~ 12 meV/kB for SmSe will close the $f \rightarrow d_f$ gap of 0.3 eV shown in Fig. 1b at a pressure of 25 kB much as observed (*5*).

[7]) $p \rightarrow f$ transitions can be identified in the actinide trihalides (*41*).

f^8; conversely, the same maximum spin alignment energy makes f^7 EuS more stable than f^6 Eu_2S_3, even though lanthanides normally prefer the 3 + state. We see in Fig. 1a that the "f-band" in EuS has fallen well below the bottom of the $(t_{2g})_f$ band (by ~ 1 – 1/2 eV), in keeping with an observed *onset* to $f \rightarrow d$ *optical* transitions of ~ 1.8 eV (see A2 and A3). Even in EuO approximately 300 kB is required to close the $f \rightarrow d$ gap and induce collapse (*5 b*). The f-band climbs so steeply with respect to the d-band that NdS is seen to be strongly confirmed as a $f^3 d^1$ material. Even the $f \rightarrow d$ optical transition energy in a dilute solid solution, SrS : Nds, should be found to have gone negative.

By considering the case of SmS, it will be seen that from Fig. 1a alone we are not in a position readily to weigh the absolute relative stability of the f^6 and $f^5 d^1$ phases. For example, why does SmS, that has been fully transformed to the $f^5 d^1$ condition under pressure, revert to the f^6 condition upon release of that pressure? Although it appears that each and every electron in the t_{2g} band in collapsed SmS is incipiently unstable with respect to capture as an f-electron, that process entrains such a large structural response, that consideration of the electronic energy in the f- and d-bands alone is not sufficient to cover developments. Any conversion of d-electrons into non-binding f-electrons gives cause to lattice expansion and a rise in the position of the d-band relative to the p-band. What is more the p-band is inferred from XPS results (*31a*) to narrow by 1 eV in this change, and probably increase a little in absolute binding energy with the increase in ionicity.

Although difficult to measure precisely by photoemission (*31*), because of charging etc., the Fermi energy in f^6 SmS would satisfyingly seem to lie at a distance above the p-band very close to that for $E_F(d)$ for the fully collapsed as it appears in Fig. 1 a, (viz. 3 – 1/2 eV). $E_F(d)$ itself was located on Fig. 1a largely through XPS from collapsed phase materials, aided to some extent by $p \rightarrow d_d$ optical assessments. f_d was next positioned so as not quite to give CeS the opportunity for further collapse to a d^2 condition (see § 1.6). This positioning moreover seems appropriately related to the XPS data on $f^{n-1} d^1$ materials, (see § 1.7).

When we turn to the second half of the lanthanide family, we see a close correspondence with the first half. Yb monochalcogenides are placed much as Eu monochalcogenides; Tm much as Sm. The *onset* energy of $f \rightarrow d$ optical transitions in these pairings (see Appendix 3) is almost identical. However under pressure the second half materials are observed to be somewhat closer to collapse implying a smaller $f \rightarrow d_f$ gap. TmTe requires only about 30 kB for full collapse, compared to 50 kB in SmTe (*5c*). TmS is actually already in the collapsed condition at STP, and TmSe has reached the ICF condition, (which we will discuss in detail shortly). In line with Figs. 4 and 5, one might attribute the above difference to the larger spin-orbit coupling of the later lanthanides. However such specific "blame" is not realistic, in view of the significant changes in binding going on across the lanthanides, such as the broadening and shifting of the d-band, as expressed in Fig. 1. Moreover f^6 Sm^{2+} and f^{13} Tm^{2+} are not complementary in the Pauli sense (being 6e vs. 1 h), and their different term arrays lead to different magnetic entropies. Such details seem to be well beyond refinement at the present stage, for closely prejudging comparative phase stabilities.

1.4 An Interpretation of how the ICF State Arises

The *f*-band used above does not of course have the characteristics of a standard one-electron band, since the *f* wavefunctions are so far from yielding metallization and the Mott transition, under intersite fluctuations of the type $f^n + f^n \rightarrow f^{n-1} + f^{n+1}$. The set complement of on-site *f*-electrons, as we have seen, reacts very strongly (~ 6 or 7 eV) in any such change. Correspondingly a system like f^6/f^7 (Sm/Eu)S is not capable of yielding rigid band behaviour[8]) (as exists through the system d^6/d^7 (Fe/Co)S_2). Only in the symmetric case of f^5/f^6 Sm_3S_4 is there to be found "heavy" intersite *f*-band hopping, which parallels the situation in Fe_3O_4.

The *f*-electrons in trivalent Sm_2S_3 have an ionization energy equal to that of some *p*-band electron, even whilst in collapsed divalent[9]) $f^5\,d^1$ SmS, as given in Fig. 1, the *f*-electron ground state level is effectively well below the Fermi level in the open *d*-band. The *f*-electrons are in general then not in a position to undergo interband particle exchange with a standard open band. Indeed the *f*-electrons are so heavy that only when they are closely degenerate with the Fermi level does this possibility arise. Because the *f*-electron band width is no more than 0.1 eV, it is a condition of some rarity. Fig. 1a points out that CeS may fall not too far from this circumstance.

A course now well known to secure its revelation is to closely investigate groups of materials which evidently already span the configurational crossover, like TmS–TmTe, or are likely to do so under reasonably attainable changes in applied conditions. TmSe has now proved an excellent example under ambient conditions of this unusual state and its fascinating properties. SmB_6 was the first material to be discovered in such a state, and SmS/Se/Te reach it, as indicated above, under relatively low pressures. The state is labelled ICF for reasons to become apparent below.

Let us consider the case of SmSe, starting from room conditions. As temporally and spatially "favored" sites are induced to release a *f*-electron into the *d*-state, and the band is rapidly freed, through occupancy, from excitonic binding and polaronic trapping, a Fermi sea of nascent *d*-electrons appears. Because the $f \rightarrow d$ conversion reduces the lattice parameters, the *d*-band is promptly required to lower so enabling the Fermi level to maintain degeneracy with the *f*-electron level. At each pressure equilibrium in the degree of conversion is sought for which the lattice parameter is compatible with continuing degeneracy between the slowly adjusting f^n level in the remaining unconverted sites and the Fermi level in the rapidly filling yet rapidly falling *d*-band. In SmSe and SmTe, experiment (*5*) shows that *all* intermediate lattice parameters can be sampled in the course of crossover, in which the Fermi level in the collapsing *d*-band and the residual f^6 levels can effectively be kept so pinned. In SmS this proves not to be possible. Lattice parameters from ~ 5.90 Å to 5.70 Å are unob-

---

[8]) The continuous linking up of the '*f*-bands' in Fig. 1 etc. between elements is for convenience only.

[9]) This and similar usage later, is made in its standard sense, in a bid to bury the misleading "valence" terminology widely adopted into R. E. jargon (e.g. "mixed valent" for ICF materials); d^1 VSe_2 or d^2 V_2O_3 are never termed pentavalent!

tainable, this material jumping over the great centre range of site conversion percentages (see Section 2). Considering the enormity of the structural and electronic changes called for with the *f*/*d* configuration crossover, it would not be surprising to find further first-order components to the transition in other materials, (which often occur at pressures where precise measurements are difficult), e.g. TmTe (*5*).

The only comparable transitions known are those involving high-spin/low-spin crossover, in for example Fe^{2+} and Fe^{3+} compounds (*11*): again cases with first-order components are found. In these *d*-electron spin transitions, (not normally involving delocalized electrons), Mossbauer results (*11*) indicate on a 10^{-9} sec. time scale, that during second-order crossover the crossover state consists of a static mixture of h.s. and l.s. sites. In the present *f*/*d* crossover, the involvement of wide band *d*-states lowers the time scale well below the 10^{-9} sec. level for the on-site interconfiguration changes (*1*). Such a rate of fluctuation leads to the highly unusual magnetic and electrical behaviour, which has brought to these systems so much attention recently (see Ref. *1*). It is these properties which earn for the crossover condition and materials caught there, the epithet "interconfiguration fluctuation", plus its acronym ICF.

In the next and later sections we shall look at how the character and condition of the *d*-state can affect the realization of the ICF condition.

1.5 The Effects on ICF State of Disorder an d-Electron Delocalization

Normally the second-order crossover is completed in a pressure span ~ 50 kB. In the case of SmS the first-order transition coming at 6 – 1/2 kB (at 300 K) precipitates the compound almost from the onset of crossover to a stage (as gauged from the observed lattice contraction) approximately 3/4 of the way through the ICF state. Possibly 30 kB more brings a completion of crossover (*5*), and a termination of the (ICF) anomalous electromagnetic properties (see App 4).

One might idly suspect that disorder would lessen first-order tendencies at crossover, particularly when that disorder is cationic. Actually this is not necessarily so, and the system (Sm/Gd) S, (which has been much examined) is found to retain a large first-order room temperature jump, skipping much of the ICF state. This suggests that anionic Te doping of TmSe might well uncover a further low pressure first-order event in the progress of reversion to f^{13}. We shall in Section 2 look at the problems of mixed crystal work in detail. We restrict its mention here to pointing out that via substitutional doping two effects of present import can be wrought upon the *d*-band. Firstly, and universally, comes the possibility of disorder induced band edge localization. Secondly, if the dopant is effectively a diluting agent (e.g. through ϵ), the width of the *d*-band can again be decreased with correlation increased. Worth investigation in the latter vein would be compounds like (Ca/Tm) S, particularly if preparable in the ordered rocksalt "111-sheeted" $NaCrS_2$-type structure (*19*). Clearly a point of some importance will be whether, in a system like (Ce/Ca) I_2 at high dilution, reverse crossover (from $f^{n-1}\,d^1$ to f^n) can be observed in circumstances that avoid the characteristic ICF behaviour, because the *d*-electrons themselves are no longer delocalized at the stage by which the interband scattering is due to occur.

It seems probable with the above model that in order to realize ICF behaviour (though not crossover itself) the *d*-band involved has to be fairly wide. For materials with inherently high-lying *f*-levels, the level may fall above the Hubbard gap in a correlation split *d*-band, and the character of the Fermi level equilibrium must then be quite different. For (undiluted) monochalcogenides there is no danger of this, it appears, and (once collapsed) the materials are quite good metals; indeed the simple d^1 metals LaS and LuS are superconductors (*12*). We emphasize though that it is a mistake to regard it as a prerequisite for crossover itself that the *d*-electron be delocalized, (witness CaF_2 : Ce, see Ref. *4*).

Unfortunately the lanthanide dihalides are very unstable, but where prepared offer revealing information on the matter of advancing *d*-band width, and the promotion of stabilization for the $d^1\,f^{n-1}$ condition. Thus in contrast to NdS and DyS, the salts $NdCl_2$ (*13*) and $DyCl_2$ (*17*) appear in the f^n condition because of the narrow *d*-band and a still tightly bound *f*-set. In the di-iodides, the *d*-band width is sufficient for collapsed phase materials to occur at the beginning of each half-period, where the "*f*-band" is highest. Indeed LaI_2, CeI_2, PrI_2 and GdI_2 are $f^{n-1}\,d^1$ metals (*14, 15*). However NdI_2 (*13*) and TmI_2 (*16*) remain as f^n insulators. The $f \rightarrow d$ optical transition energy is, we suspect, virtually identical between a sulphide and an iodide, so thet noncollapse of NdI_2 and TmI_2 we attribute directly to the more sensitive loss of *d*-band width. In NdI_2 this may be abetted by its adoption of the awkward 7-fold coordinated SrI_2-type structure (*17*), rather than CdI_2-type. Pressure studies on both compounds are worth pursuing. TmI_2, with its simple CdI_2 structure, will in all probability support *d*-band delocalization, if collapse can be induced, so that ICF behaviour should again be encountered. However with such a material the appearance during the crossover of metallic properties might be delayed, for it may prove not possible for electrons to be released properly into the *d*-band to begin with. Heavy *d*-electrons might initially be unable to sense the full *d*-band width, restricted as they may be to the reverse hopping of an *f*-electron in inter-site migration. This effect could introduce a cooperative first-order component to crossover.

When discussing mixed crystals in § 2, we shall return once again to the role the effective *d*-band width plays in promoting collapse and ICF conditions.

1.6 The Effects of a High-Rising f-Band in Ce and Tb Compounds

We turn now to the *front* end of each half-series, and the divalent[10]) compounds of Ce and Tb. Fig. 1a makes clear that the configurational choice offered here is not

[10]) ibid, pg. 15.

f^n vs $f^{n-1} d^1$, but $f^{n-1} d^1$ vs $f^{n-2} d^2$; i.e. for Ce^{2+}, $f^1 d^1$ vs d^2. Actually CeS/Se/Te are found to be magnetic $f^1 d^1$ metals (*20*), although Ce metal is known under pressure to take up the configuration $d^2(s^2)$ (*21*); (only the free Ce^{2+} ion is f^2). The $4f^1\ 5d^1$ condition of CeS, etc. contrasts with that of the corresponding actinide chalcogenides, ThX (*22*), which are non-magnetic $5f^0\ 6d^2$ metals, thereby demonstrating the looser 5*f*-state binding. Though magnetic behaviour is not recorded, CeI_2 is presumed to be $f^1 d^1$. ThI_2 is known (*23*) once again to be $5f^0\ 6d^2$. With quadrivalent materials like US_2 (*24*), even 5*f* binding is so greatly augmented that the $5f^2$ configuration is now strongly held.

Turning to trivalent compounds, possibilities for ambiguous behaviour once more are afforded to Ce and Tb; c.f. CeN/P, TbN. In this trivalent case, the prerequisites here are a sufficiently high-reaching *f*-band position in the expanded state, plus a sufficiently low dipping *d*-band there. For $3d^0$ ScP the $p \rightarrow d$ band-gap is 1.1 eV (*25*), so that in a RE phosphide the $p \rightarrow d$ gap probably is ~ 1.5 eV, this diminishing steadily towards zero in compounds like CeAs/Sb. In CeN (and CeP under pressure) (*26*) one actually does obtain a substantially[11]) collapsed $f^0 d^1$ metal. However, with the arsenide and antimonide, the *f*-electron apparently sinks into the *p*-band, unretarded, like the *d*-function, by extensive *p*-hybridization, and the $f^1 d^0$ condition returns[12]). We say "returns" here, because adherence to the f^1 choice is also to be found in narrow *d*-band $CeCl_3$, etc. There the *d*-band bottom is too high to receive the electrons, and moreover would be Mott insulating if it were to. For the potentially interesting case of CeI_3, its $PuBr_3$-type structure (*27*), (close to d^0 ZrS_3), probably will pre-empt the appearance of any obvious delocalization-cum-configurational change, because of the possibility this structure affords for additional covalent pair-bonding, (c.f. semiconducting d^1 TaS_3) (*28*). ThI_3 probably is already in this latter condition. The magnetic properties of neither iodide are known. Much remains to be explored for these various halides; as a class however they are very hygroscopic (*27*).

1.7 Photoemission, Relaxation, and the f-Band Position

It was seen just above that the *f*-electron in trivalent f^1 CeSb must be fairly close to being degenerate with the bottom of the (expanded phase) *d*-band, which is itself degenerate with the top of the *p*-band. X-ray photoemission (*29*) from this *f*-state however would indicate a binding energy that is tighter by 2 to 3 eV. (The complete set of antimonide XPS *f*-shell ionization results is to be found incorporated in Fig. 1 a.) We take this augmentation in excitation energy to reflect the relaxation of the effective eigenstate over the $f^1 \rightarrow f^0$ ionization process. Following *Slater* (*30*), the

[11]) The condition has not yet been examined in detail.

[12]) The semimetallic nature of these f^n compounds is due either to poor stoichiometry, or more probably in the antimonides to slight *p*/*d* overlap.

observed ionization energy relates to some intermediate value between the initial and final one-electron eigen-values for the f-band. A rather similar, though probably somewhat smaller relaxation to the above, is to be expected when one of the f^n-electrons is photoemitted from divalent expanded SmS or EuS (*31*). Experimentally the first f-state XPS peak, (i.e. the one associated with a final state left in its ground-state many-electron coupling) falls respectively in these materials at + 2 – 1/2 and + 1 eV (above the top of the p-band), and indeed these values are some 1 – 1/2 eV below the undisturbed f-band positions as proscribed by Fig. 1a. Next we shall claim that, accompanying f-electron XPS from the collapsed phase $f^{n-1}\,d^1$ materials (*31*), a considerably stronger relaxation of the ionizing f-band occurs of about 4 – 1/2 eV, so that with divalent $f^{n-1}\,d^1$ materials it appears almost as if we were ionizing the corresponding f^{n-1} trivalent cation. This arises because the f-electrons of $4f^{n-1}\,5d^1$ are significantly separated, spatially if not energetically, from the outer d-electron. In the course of added depletion of their number by ionization, an excessive tightening of all inner f-functions is now encountered, since in the $f^{n-1}\,d^1$ case the outlying d-electron is unable to participate at all equitably in any shielding of the ionizing f-electron from the incurred nuclear charge imbalance. Fig. 1a again shows these results.

We shall close this section by requesting again that our incorporation of the f-electrons into a one-electron band picture be treated with understanding. Only under rare circumstances, (namely those which bring about ICF conditions – when the Fermi level for the growing d-electron sea and the old f-electron level in the remaining f^n sites become temporarily degenerate in the mixed system), are the f-electrons able to become active at the Fermi energy, and a transient equilibrated interchange with the d-electron sea to occur. The phenomenon is enountered only very occasionally, as for SmSe between ~ 5 and 50 kB. The condition is one encountered "incidently" en route between two isostructural "electronic polymorphs". The transition involves a large change in volume, and much change in relative position of the tight-binding "bands". For the f-band we show a shift of 2 – 1/2 eV. It is the large response of the f-band to configuration and occupancy changes which allows the "f-band" to sit so far below the d-band-determined Fermi level in collapsed TmS etc., and the p-band-determined Fermi level in TmSb etc. Only in rare cases as with hypothetical f^6 EuSb or f^{13} YbSb is instability encountered with respect to spontaneous reduction, f^7 Eu^{2+} and f^{14} Yb^{2+} alone being capable of holding a further electron below the top of the p-band. In transition metal compounds the position is much less extreme, disproportionation of d^9 CuI_2 to Cu and d^{10} CuI being typical. The partial autoreduction in "CuS" is a further example (*2*). For an overwhelming percentage of the time, even in ICF behaviour, each unit cell has some integral complement of f-electrons because the f-wavefunctions are so localized. Likewise the f-states in a lanthanide solid solution are never rapidly time-averaged in rigid-band fashion, as in many a d-band system, such as (Ti/Ta)S_2. In the next section we shall see something of what does happen in mixed rare-earth systems.

2. Comments on the Mixed Systems. The Mott Transition and Disorder

2.1 Dopant Size Effects and ICF Equilibrium Conditions

Much light is shed on questions like the above by results recently obtained with mixed cation systems of the type (Sm/Y)S, (Sm/Gd)S, etc. (*32, 31b, 7*). Such systems have been much investigated, as they are found to lead to the ICF condition in SmS without resorting to pressure. The systems raise the interesting question of to what extent d^1 doping of SmS truly mimics the effect of pressure.

Most d^1 RE sulphides (even LaS) are smaller than expanded SmS (a_0 = 5.96 Å), and so, when it was first found that such d^1 doping collapses SmS, the action was often attributed specifically to the lattice parameter change. This was encouraged by facts such as YS (a_0 = 5.49 Å) producing collapse at only half the concentration (15%) required by the larger, but chemically similar, LaS (a_0 = 5.85 Å). However the existence of an added factor, beyond simply the relative sizes in a_0 has become apparent, connected with the possession of delocalized *d*-electrons. Thus d^2 ThS, where a_0 = 5.68 Å, produces collapse in SmS even faster than does d^1 GdS, where a_0 is down at 5.56 Å. Conversely, d^0 CaS (a_0 again 5.68 Å) does not succeed, (and neither in all probability would MgS (a_0 only 5.19 Å) or d^5 MnS (a_0 = 5.23 Å)), in collapsing their SmS host at *any* concentration. We shall expand in § 2.4 on this added factor.

In Fig. 2a we indicate the base-lines which we have used in analyzing the lattice parameter data from Ref. (*5*) on the (Sm/Gd)S system. Values for the hypothetical materials f^8 GdS and (room pressure) fully collapsed $f^5 d^1$ SmS, were extrapolated from Fig. 1 of Ref. (*33*). SmS at STP, we determine, via the tangential technique indicated in Fig. 2a, to be very slightly not pure f^6. This is of course under the presumption of no negative departure from Vegard's Law. The results of the analysis are presented in Figs. 2b and 2c. There we show the overall percentage of d^1 in the mixture, and, more revealingly, the percentage of Sm content in the mixture that stands "converted" to (i.e., which has fluctuated into at any instant) the $f^5 d^1$ configuration under the given conditions. This latter percentage blows up as the crucial room temperature content of 16% Gd[13]) is approached. Yttrium doping appears very similar (*32a*). Interestingly the above critical percentage is that for which the maximum is reached in the number of Sm with two nearest neighbour Gd, (and moreover, that for which the rate of growth of those with 4 *nn* Gd is greatest). From Fig. 2c (*2*) it further seems to be the point, where the Gd finally manages to induce (at each instant) its own number of $f^5 d^1$ Sm sites. How general this latter result is for other mixed systems, and to what extent it is modified by a_0 for the d^1 dopant will be interesting to follow up, (e.g. for (Sm/Lu)S; a_0 for LuS is only 5.35 Å). It does, however, seem quite general that at low concentrations, ($x \lesssim 10\%$), the dopant is "inefficient" in promoting Sm conversion. Only when there becomes an appreciable number of Sm

[13]) This percentage is only slightly dependent on temperature (see Fig. 2b) for T > 100 K.

with three and more n.n. Gd does the system start to shift rapidly towards collapse[14]). This same feature is further mirrored by the form of the non-linearity (– a very strong upward bow) in the plot of $P_{300\,K}$ vs. x_{Gd} for the conditions relating to the B → M transition in mixed crystals (see Ref. *5a*, Fig. 6).

It should be cautioned here not to view the above conversion percentages derived from Fig. 2a, too rigidly. Firstly, they depend on assuming a Vegard's law relation between two electronically dissimilar end materials (*34*). A downward bow in the *base-line* would augment the percentage of f^6 SmS throughout. Secondly, application of Vegard's law presupposes that, for example, the ionic radius of f^6 Sm^{2+} is intrinsically unchanging as the doping brings the crossover closer. However, it is most likely here that the ion size decreases somewhat, even while still holding to the "f^6" configuration. The *d*-electron wavefunctions from collapsed sites will for one thing penetrate neighbour anion sites by covalent admixing, and some *f*/*d* hybridization results. Further, it is not known precisely how long any unit cell might take in which to relax upon its cation returning from a $f^5\,d^1$ excursion. In view of such prospects, it is not impossible that just prior to collapse, the percentage of the Sm sites standing converted actually could be $< 10\%$.

It would seem feasible that sites prone to making frequent and/or extended excursions into the $f^5\,d^1$ state are those with, by chance, three or more d^1 nearest neighbour sites (inc. collapsed Sm). However, from the observed inhibition of f^5 $^6H_{5/2}$ magnetic susceptibility (Ref. *32c*, Fig. 3), it is taken that the vast majority of such sites are *not* in fact frozen in the collapsed configuration, but do actively participate in configuration fluctuation. In fact, Sm Mossbauer isomer shift measurements (*35*), by always yielding just a single line, indicate that *all* site types (fractional representation f_s) undergo fluctuation at rates of not less than 10^9 sec^{-1}. One can here presume though that less heavily neighboured Sm sites will exist for a smaller fraction (ϕ_s) of their time in the $f^5\,d^1$ configuration. Then, for the system as a whole –

$$\sum_{\text{types of site}}^{s} f_s \phi_s = \Phi_d\,,$$

the overall fraction of Sm sites instantaneously converted to $f^5\,d^1$, as recorded approximately by Figs. 2b/c. Unfortunately Mossbauer measurements have not been made yet for mixed systems other than $Sm_{0.77}Y_{0.23}S$; moreover even this material was not taken into the low temperature B′ phase, (see below).

[14]) There does not seem to be entailed here any discontinuity at the dopant's Mott transition ($x_{Gd} \approx 10\%$) – see § 2.2.

2.2 Electrical Behavior. The B′ Phase an Expression of Disorder

Earlier when discussing binary R.E. materials, it was observed that for the interconfiguration fluctuation phenomena to arise customarily involved a material, upon gaining the $f^{n-1}\,d^1$ configuration to be through its Mott (-Anderson) transition. Interestingly enough, for the mixed systems like (Sm/Nd)Se (*7*), or (Sm/Y)S (*32b*), there is no indication of the host being precipitated into collapse upon that critical concentration of the dopant being reached for which the *latter* (disordered) array is carried through *its* delocalization transition. In chalcogenide systems like the above, this event relating to the *dopant* subarray seems (at 300 K) to occur at $x \approx 0.10$ (*36*), whilst, for the particular case of (Sm/Gd) S, the precipitous collapse at the host sites does not in fact come until $x \approx 0.15$. A more definitive view of the earlier process should be obtained with a system like (Lu/Yb) S, or even (La/Sr) S, if preparable[15]). Concerning the latter conditions, governing and relating to rapid configuration crossover in the *host*, insight recently gained from the transport properties of these mixed systems proves them to be more complex than anticipated.

With x beyond the dopant's Mott transition *but* short of "collapsing" the host, Hall measurements actually reveal that, however many of the Sm sites may then stand instantaneously converted to the $f^5\,d^1$ configuration, these same sites are constrained not to release their *d*-electrons into the Fermi sea. Furthermore, at low temperatures the above non-release of *d*-electrons from converted host sites continues to be true to much larger x, over the phase region termed B′ in Fig. 2b. Given a sample in this phase region, release and a first order collapse into the metallic ICF phase (M) is secured upon warming[16]). With this transition the Sm is brought more fully to take up the $f^5\,d^1$ configuration, and the ICF equilibrium is shifted sufficiently as to incur the restoration of overall metallic conditions[17]) much as prevail at 300 K (for $x \gtrsim 16\%$). Upon going as above from B′ to M, the Hall coefficient duly registers a substantial increase in free carrier density, – although there is noted to be a strange reversal in its sign.

As with pure SmS under pressure (*37*), the above metallized ICF conditions yield (*9c*) a free-carrier plasma edge which, though reaching well into the visible, is found to be restrained by ~ 1/2 eV, compared with simple d^1 metals like YS (*9c*) or LaS (or $f^7\,d^1$ GdS (*12a*) also). In line with this observation, and even more striking, is that the nett d.c. *resistivity* is actually increased upon going from B′ to M. Clearly, after the Sm $f^5\,d^1$ sites finally do become joined in *d*-band activities, the continual annihilation and creation of carriers at the Sm sites constitutes a strong scattering

---

15) We presume here in the (Lu/Yb) S case that the Yb sites will be almost exclusively $f^{14}\,d^0$, i.e. that the Lu d^1 electrons will not readily penetrate these sites because of the d_f/d_d energy difference. In (La/Sr) S this mismatch is even larger.

16) Thermal cycling across this transition completely shatters the crystal. This is presumably due to doping inhomogeneity on a fine scale.

17) i.e. all sites will in the course of time suffer metallic *d*-occupancy fluctuations, – besides all Sm sites suffering the six electron configurational activity.

potential. This is so not only via the continued temporally and spatially fluctuating strain field experienced by all the carriers, but now also via individual carriers directly involved in such processes. The conditions are reminiscent of, though clearly more involved than those which prevail in an amorphous metal. There is a strong emphasis on the local short-range aspect.

The interloping form taken by the low temperature B′ phase on the $T - x$ phase diagram (Fig. 2b) would indeed support that its origin, as is implied above, lies in disorder localization of the Sm originating *d*-band carriers. The disorder is able to sustain a situation at low temperatures in which the Sm (over its $f^5 d^1$ sites) is not through its Mott-Anderson transition until a point much beyond where the dopant is through its. This then permits a very high conversion percentage of Sm to be attained without bringing about the first-order collapse.

In a mixed system like (Sm/Gd)S we have seen that, in phase region B′, the Sm d^1 electrons are excluded from the Fermi sea. Metallic conduction persists essentially using only the electrons of the Gd subarray with the corresponding wavefunctions tending to avoid the Sm sites. The resulting metal though not a good one does show a *positive* temperature coefficient. This same feature is found also for the mixed systems in the M phase at all temperatures, and in this there arises a marked contrast to the behaviour of a *pure* material in its metallic ICF condition.

2.3 The Influence of Configurational Disorder on the M Phase of Pure Materials

One of the striking properties first to draw attention to ICF "metals" (like SmB_6, SmS under pressure, or TmSe) indeed was their steadily *in*creasing resistivity upon cooling below 200 K. Clearly in the mixed systems discussed above, the dopant array manages to short out the conduction difficulties that are developing on the host array at low temperatures. To date it has not been too closely pursued as to whether the B′ phase finally does put in an appearance on the T – P phase diagram of pure materials at low temperatures. In light of the fact that much of the randomness in the mixed systems is actually an intrinsic feature of the ICF condition itself, over and above the alloying, a careful search for this phase in SmS under pressure should not be a misguided effort. The mixed anion system Tm(Se/Te) may however provide a more encouraging starter. Actually in the original low temperature work on SmS under pressure (*38*), it did appear that slightly greater pressure was required to secure "collapse" at 4 K[18]) than at 473 K, in conflict with the positive slope to dT/dP (now confirmed down as far as 100 K (*5a*)) for the boundary separating the B and M phases. Magnetic and lattice entropies will however mitigate against the appearance of B′ at all but the lowest temperatures (see Ref. *5a*). Implicit from the slope of the B′ → M

[18]) The transition as recorded could also have become second order.

phase boundary is that ΔS must be small and positive, in contrast to being large and negative across the B $\rightarrow$ M boundary. It is to be expected that B′ should have the lowest entropy and enthalpy of the three phases.

The form taken by the resistivity rise at low temperatures in the ICF metallic phase has not yet been satisfactorily analyzed. Fig. 3 shows that a $\rho \propto \exp(T^{-1/4})$ relation makes fair representation of the data (*38*) for SmS at 20 kB. For 10 kB, SmS is not "properly" converted at the lowest temperatures (Ref. *38*, inset to Fig. 2) and the sigmoid deviation shown could be ascribed to this cause. However none of the data presently available for ICF metals (e.g. SmB_6, TmSe), seems to express an exponential run away in resistivity to low temperatures. It is possible that the observed flattening off marks the limit set by surface conduction, via permanently and fully

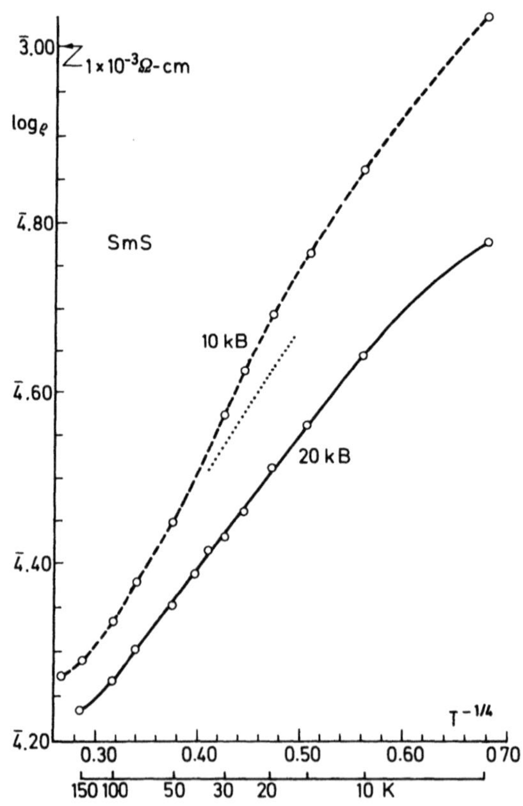

Fig. 3. ln ρ vs $T^{-1/4}$ plot below 150 K of the data from Ref. (*38*) on SmS at 20 kB; this is well into the metallic ICF phase (M), unlike 10 kB, which possibly catchs a B′-like phase around 40 K. In the latter phase, of course, there would be no interspersed permanently d^1 sites, unlike (Sm/Y) S, etc.

collapsed surface "shorts". The anisotropic environment of crystal edges and steps may be sufficient to produce such pathways. In the present complex electronic circumstances, moreover, it is not really to be expected that a simple phonon assisted hopping $T^{-1/4}$ law should hold. For the equally complex case of "disorder localization" in the CDW system 1 T – $Fe_xTa_{1-x}S_2$ (*42*), the data are best fit empirically by a power law, with the concentration x in the exponent! The disordering Fe^{2+} t_{2g}^6 sites here sit outside (i.e. do not contribute electrons to) the CDW-supporting conduction band formed over the d^1 Ta^{4+} and iron-compensating d^0 Ta^{5+} sites.

An increase in R_H under disorder localization seems quite normal, and is known to be present in SmB_6 below 50 K (*43*). Similarly it is not surprising to find that the low temperature electronic specific heat in metallic ICF materials is very high ($\gamma \sim 150$ mJ/mole K^2 in SmB_6 or h.p. SmS (*38*), – while much lower of course in LaB_6 or LaS). Very little exploration of mixed narrow-band *d*-metals – say d^1 (La/Lu)S or d^2 (Mo/W) O_2 – has been undertaken to know what rise in γ might be expected in such metals from straight cation disorder. For the system Ti_2O_3 doped to metallic concentrations with non-magnetic d^2 V^{3+}, in which strong modifications in local bonding and lattice dimensions are incurred as a carrier migrates, it is known that γ can rise to an extremely high value (*39*). This lattice effect is distinct from the strictly electronic effect that enhances γ in a highly correlated metal like V_2O_3 on the borderline of the Mott transition. Needless to say γ is still further enhanced in such a metal by cation substitutional doping, even in cases when, from the electronic point of view, that doping would take the material further away from the Mott transition, – as for example Ti in V_2O_3 (*40*). Some further points concerning pure SmS, which are best deferred until after § 2.4, have been collected into Appendix 4.

2.4 Disorder and the d-Electron Band in Cation Doping

We now return finally to take up the point with which we ended the first paragraph in section 2.1; viz. via what means is doping that bears delocalized *d*-electrons successful in promoting configuration crossover in a solid like SmS or TmTe? Earlier we illustrated, by citing that CaS (a_0 = 5.68 Å) doping of SmS (a_0 = 5.96 Å) leaves the f^6 state unchanged, that the effect is not a straight mimicing, through the reduced lattice parameter, of the effect of pressure. Indeed we now believe, (though this has not yet been tested), that even if the d^1 doping should entail a slight expansion, as for La in TmTe, the effects from these delocalized *d*-electrons would still be sufficiently potent as to lead to "collapse".

Only when the *d*-band is sufficiently well formed, running over all sites, host and dopant alike, is the full benefit of configuration crossover attained. We have seen, with the B′ phase of (Sm/Gd)S, that when the Sm f^5 d^1 sites are excluded from the Fermi sea, the ICF conversion percentage becomes relatively low. This is so despite the continuing metallic background of the dopant d^1 gas. With a dopant that is unable to promote metallic conditions the position will be worse.

Let us try to unravel some of the effects that can be introduced by cation doping, and see how it works out in practice. If we dope say SmS with CaS, there will be an effective pressure on the Sm sites that tends to raise the f^6 level. However the f to t_{2g} gap will not be reduced should the empty d-band recede at a faster rate. This we envisage because the d-band in CaS is high-lying. Still smaller MgS probably will not succeed either, because it carries no d-function to form a rigid-band, and hence the d_f-band narrows simply because of dilution. All such cation mixing moreover disorders the bottom of the d_f-band, so retarding release of metallic electrons. This touchs on we believe, another key function served by d^1 doping; namely to raise the Fermi level away from the 'roughened' edge of the d-band so that electron release from an f^n site is not held back by Mott-Anderson localization. The most important function which d^1 doping serves, however, in promoting "collapse", is neither to raise f^n nor to raise E_F from the d-band edge, but very rapidly to lower that d-band onto the f^n level. We have seen in § 1.1 that the d^1 occupied d-band in the lanthanides lies 2 eV below the unoccupied band (d_f). In a mixture like (Sm/Gd)S we presume that rigid d-band behaviour is fairly readily secured, and by only ~ 10% Gd, the Fermi level in the falling d-band is degenerate with the f^6 level. A higher concentration of La is required (*44*) for this because the d-band in LaS is somewhat higher than in GdS. In the system (Tm/La)Te montioned in § 2.1, collapse should still be secured under this fall in the d-band position, even though the lattice parameter of LaTe is greater than TmTe. Th-doping secures collapse of SmS even sooner than does Gd-doping, as the d^2 occupancy and the increased principal quantum number probably unite to lower the d-band in ThS relative to the p-band[19]). YS is we surmise better than LaS in promoting collapse because of its smaller size. It yields a system much as (Sm/Gd) S and unlike (Sm/La) S which has lost the strong first order component. LuS should be more effective even than GdS, (unless the d-band is destabilized somewhat as a result of a contraction of the now closed f^{14} shell). $f^{14}\,d^0$ YbS should on all counts cause TmS to become f^{13}. As for (Tm/Sm) S the system should provide a fascinating conflict: probably TmS will collapse SmS before SmS expands TmS, but the same question for the selenides is more in doubt. Mixed selenides have not been much examined. The crossover in the system (Sm/Th) Se probably will remain second-order, and occur at lower doping content than the second-order crossover in (Sm/La) S (~ 30%) (*44*). It would give clearer lattice parameter information on the course of the crossover through ICF conditions than does the latter system, $Th^{2+}\,d^2$ being smaller than $La^{2+}\,d^1$. Sr-doping of TmSe could bring about reversion to the f^{13} condition also in second-order fashion.

Iso-electronic *anionic* substitution probably has the effect of delaying collapse somewhat in a system like Sm(S/Se), relative to a linear interpolation, because of some disorder at the bottom of the unified d_f-band. This is likely to be less severe though than in a mixed cation system like (Sm/Yb) S. The effect of *non*-isoelectronic

---

19) A lower, broader (i.e. more correlation free) d-band in ThS than GdS should be confirmed by XPS measurements.

anionic substitution, as in Sm(S/As), is more complex. It is found that As-doping of SmS rapidly leads to a first-order transition that carries the ICF equilibrium well towards full $f^5\,d^1$. This is outwardly much as for collapse under pressure, and indeed the origin for removal of the $f \rightarrow d$ gap is probably not too different. As-doping replaces a small S^{2-} anion by a large As^{3-}, so that the remaining f^6 Sm^{2+} sites react as to a large local pressure. The collapse is not strongly coupled to the presence of Sm^{3+} f^5 sites, as can be seen be the non-collapsed condition (*45*) of the Th_3P_4-type double sulphide Sm_3S_4 (i.e. $Sm_2^{3+}Sm^{2+}S_4$). This material is probably less close to collapse than SmS, because of a higher unified *d*-band, and the effects of cation disorder. These features do not prevent the Nd^{2+} in Nd_3S_4 from being $f^3\,d^1$, following from the much higher-lying position of f^4 than f^6 on Fig. 1.

2.5 Endpiece

It is evident there are many points yet to be understood even qualitatively in this field, and a flood of experiments still suggest themselves, only a few of which have been mentioned here.

With this work an effort has been made to relate and reduce a wide range of facts and features to a closely specified one-electron band scheme, accommodating as fas as possible the demands made by correlation, disorder, etc. Although much of the basic material has been extensively discussed for some time, it is hoped that the subject does with this paper become further clarified and consolidated.

Any privileges due to a hitherto "outsider" to R.E. work, I would be glad to claim concerning referencing, etc. The work begins and ends as a rather personalized essay in understanding, and was not in any way intended to be a full review of work to date on these topics.

Thanks are due to several colleagues who have read through parts or all of the manuscript and offered their constructive comments, especially *W. F. Brinkman, M. Campagna, A. Jayaraman, C. Varma* and *G. K. Wertheim.*

Appendices

After reading Section 1 several colleagues have suggested amplifying the background to Fig. 1, in particular as it concerns the f-states. Unfortunately Professor Jørgensen's work although so prolific, is not as well known among solid-state physicists as it ought to be. I can do little more particularly in Appendix 1 than to give my understanding of that work. Its most recent addition is a review to appear in The Gmelin Handbook of Inorganic Chemistry, Vol. 39 (B1), on "Rare Earths" (1976).

(A1) The Origin of the Shape Given to the f-band in Fig. 1.

We are interested in the ground-state term of each configuration f^n, and its distance below some fixed reference level. The first step is to obtain the barycentre of all states of the given configuration having the S_{max}, Hund's rule, condition, and to determine its stabilization with respect to the barycentre of all states of the configuration f^n. Secondly, the movement of the grand barycentre needs to be established on an absolute energy scale, as a function of n. It has been determined in fact that both the total spin alignment energy[20] and the binding energy of the grand f^n barycentre follow the form $\frac{n}{2}(n-1)\epsilon'$, appropriate to particulate interaction avoiding self-energy and double-counting. The grand barycentre shift between one element and the next takes the empiric augmentation n (E–A) = 0.4 eV. The form "E–A" used here for the incremental energy unit, ϵ', expresses not only increase in central binding (from the increase in Z) but also the countering increase in interelectronic Coulomb repulsion (from the increase in n). The parabolas relating to both above barycentres are shown in Fig. 4, for conditions appropriate to $M^{3+}(f^n)$. Their sum is given also (curve c).

Since we are actually interested not just in S_{max} and its barycentre, but in the lowest S, L term, and moreover in its lowest J-level, the above curve must now be "corrected". The first (and somewhat larger) correction comes, when, for L = 5 and 6, S_{max} is able to be produced for values of L other than just L_{max}. The secondary correction involving the spin-orbit coupling stabilization of the lowest[21] J-value is largest also when L is largest, although it is never greater than 1 eV. These 'corrections' relating to the ground-state term have been incorporated to yield the fourth and final curve on Fig. 4.

As derived from Fig. 4, we plot next in Fig. 5 the *augmentation* in the f^n ground-state energy (relative to the external reference) which arises between successive elements. This is what in Fig. 1 was made to serve as one-electron-like energy for the nth electron of configuration f^n, to describe processes involving f-electron loss such as photoionization, or intracation $f \rightarrow d$ transitions. This seems feasible because both the p- and d-bands and the work function are all relatively invariant in energy across any particular series of lanthanide compounds.

Figure 6 shows the observed $f \rightarrow d$ excitation energies, mentioned in § 1.1, for M^{3+} and M^{2+} in CaF_2. The detailed aspect of the two sets of data is seen to be almost independent of the valence change, aside, that is, from the rigid shift of 6 – 1/4 eV (50 kK) between the two.

We see then that the 'corrections' made above are what give to such plots their characteristic shape. Proceeding without them would give two straight lines. In Fig. 1 the energy scaling is contracted somewhat compared with Fig. 6, particularly for $f^n d^0$, to match the concentrated cation situation (see App. A2).

---

20) reversing beyond $n = 7$.

21) highest beyond $n = 7$.

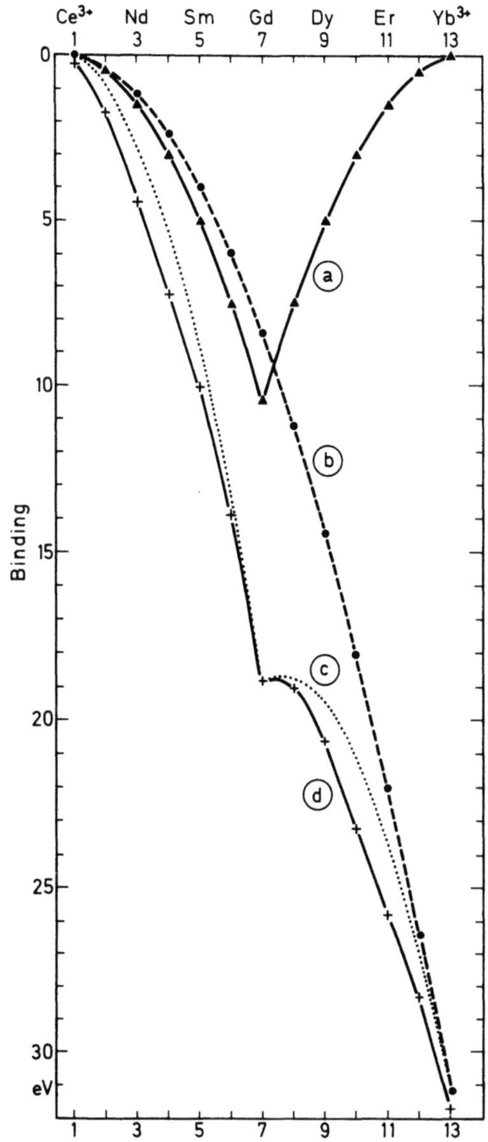

Fig. 4. Concerning a one-electron representation of the *f*-electron energy in configuration f^n ground-state. Numbers used in theoretical expressions appropriate to Ln^{3+}.

Curve *a* – spin-pairing energy of f^n set; i.e. S_{max} barycentre relative to *b*.

Curve *b* – barycentre of all states of f^n set, relative to "outside".

Curve *c* – sum of *a* and *b*.

Curve *d* – as *c*, but "corrected" to most stable level with S_{max}, giving energy of ground-state of f^n set below external reference. Latter, within compound series, approximately fixed at appropriate level w.r.t. *p*- and *d*-bands.

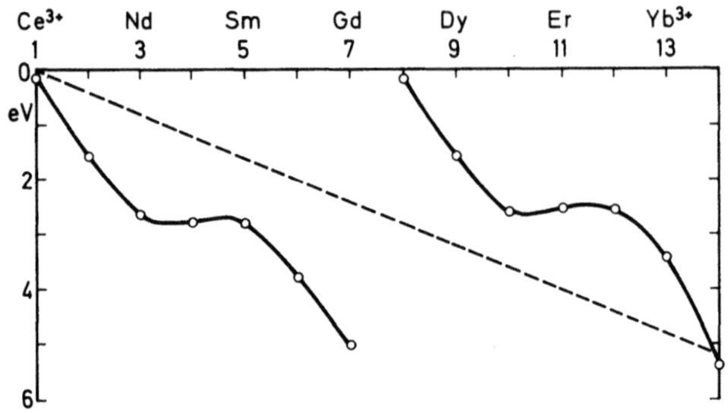

Fig. 5. Incremental energy, from curve *d* of Fig. 4, on passing from element to element. Considered as one-electron energy of the additional electron for use on Fig. 1. The dotted line shows similar incremental augmentation in binding relating to barycentre of all f^n coupled states across lanthanide series.

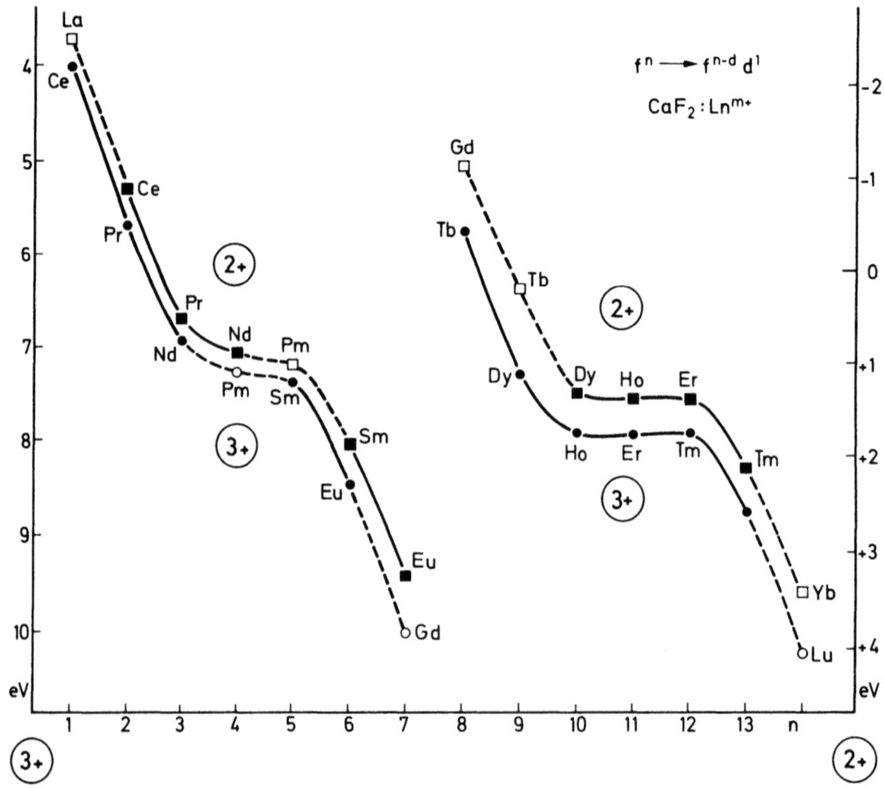

Fig. 6. Experimental values for lowest $f \rightarrow d$ optical transitions in cases of CaF_2 doped with Ln^{2+} and with Ln^{3+}. The only difference between the two is a 6 eV augmentation for the latter.

(A2) Nephelauxetic Effect and Reversal of $f \rightarrow d$ Excitation Energy

Figure 7 should serve to illustrate the text, concerning our opinion of how this reversal comes about. For terms internal to the f^n set, there occurs only a relatively small nephelauxetic reduction in excitation energies with materials of advancing covalency. However the *differential* effects between the f and d levels are, it is evident, quite large, and not monotonic. Advanced p/d mixing inhibits further expansion of the f wavefunctions. How the energy of the final state optically excited d-electron should be correlated with the bands on Fig. 1 is dealt with in the main text.

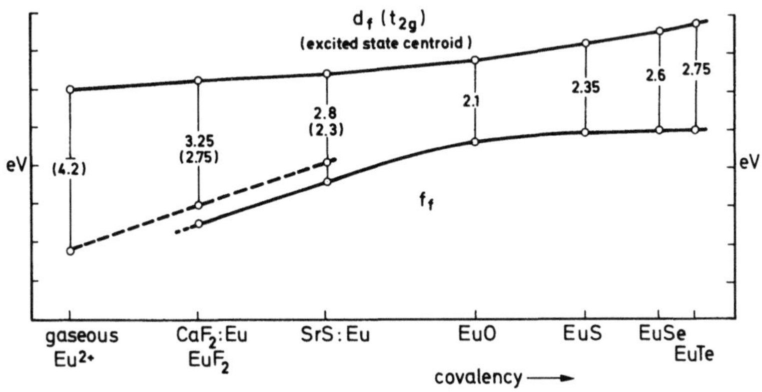

Fig. 7. Observed values for sequence of Eu^{2+} materials of first $f \rightarrow d$ optical transition. An attempt is made to show the behaviour of the t_{2g} state (terminal) w.r.t. the f-state during the nephelauxetic changes associated with the advance of covalency. The dotted line refers to the state J = 0, absolute ground state of the f^6 residue. In the other materials the individual J lines (7) are not resolved. (Close multipleting is also encountered for Sm^{2+}, but not Tm^{2+} or Yb^{2+}.)

(A3) Sources and Interpretation of Optical and XPS Data for Fig. 1

In the following tables we try to show the data considered in the construction of Fig.1. Data on the selenides and tellurides is included to widen the base for interpretation. It is quickly evident that this data base is still very sketchy. In part this is due to the very broad featured results that come from the $f^{n-1} d^1$ metals, but even where sharp spectra could be obtained there are still many gaps, eg. LuS and TmS, XPS; YbS, optical and XPS. [The f2 etc. denote figure numbers in the references indicated; auxiliary references rAb etc. can be found after the references to the main text].

i) $p \rightarrow s$ $(\Gamma \rightarrow x)$

	Y	La	Sm	Gd	(eV) Eu	Tm	Yb
S	5.5 (r9c, f2)	broad (rAe, f2)	5.5 (r9e, f3)	broad (rAe, f2)	5.0 (rAb, f1) (rAf, f5)		
Se				broad (rAe, f2)	4.7 (rAf, f6)		
Te				broad (rAe, f2)	4.4 (rAf, f6)	4.6 (rAg, f2)	4.9 (rAd, f1)

(rAa)	Ba	Sr	Ca
S	3.9	4.8	5.4
Se	3.6	4.5	4.9
Te	3.1	3.8	4.3 eV

ii) $p \rightarrow d_f$

	Sm	Eu	Tm	Yb
S	~ 4 (r9c, f1)	4.0 (rAf, f8)	–	
Se		3.7 (rAf, f8)		3.6 (rAc, f1)
Te		3.3 (rAf, f8)	3.5 (rAg, f3) (rAh, f1)	3.6 (rAc, f1) (rAd, f1)

iii) $p \rightarrow d_d (E_F)$

YS	4.2 eV	(r9c, f1)
LaS, GdS	~ 4	(rAe, f3)
TmS	~ 4	(rAh, f1)
TmSe	~ 3.5	(rAh, f1)

iv) f_d *XPS*

YS spectrum shows $p \rightarrow E_F \sim 4.2$ eV, d-band filling ~ 1.5 eV, (r44, f1a).

NdS, GdS, DyS, ErS, $p \rightarrow E_F \sim 3.5$ eV, and f emission excited hole analysis, (r31a, f1).

- resolution only fair here, see SmS vs r31b.
- line identification in DyS 2 eV too low.
- p-band width in f_d materials seen to be ~ 1 eV wider than in f_f.

v) f_f *XPS*

First number: first *f*-peak, relative to *p*-band.
Second number: ditto, relative to E_F, (susceptible to charging, impurities, etc).

	Sm	Eu	Tm
S	2.5 (0.8) (r31b, f1)	1.0 (2.8) (r31a, f1)	–
Se	2.0 (1.1) (r31b, f1)		2.2 (0.5)[a]) (r18, f2)
Te	1.5 (1.4) (r31b, f1)		– (0.8)[a]) (r18, f2)

[a]) Surface conditions augment $f^{n-1}d'$ concentration.

vi) *Antimonide, XPS* (r29)

E_F just at top of *p*-band, due to slight *p*/*d* overlap. CeSb shows f^1 ionization to f^0.

vii) $f \rightarrow d_f$

Optical data: leaving f^{n-1} residue in ground state L, S condition; J-averaged in Sm and Eu cases.

	Sm^{2+}	Eu^{2+}	Tm^{2+}	Yb^{2+}
f^{n-1} residue.	f^5	f^6	f^{12}	f^{13}
Its ground state.	$^6H_{5/2}$	$^7F_0$	$^3H_6$	$^2F_{7/2}$
levels closer than 1 eV.	8	6	1	0

Excited state map given in "Physics of Rare Earth Solids" by K.N.R. Taylor and M.I. Darby (Chapman and Hall, 1972), p 10.

	Sm	Eu	Tm	Yb
S	0.8 (r9e, f1)	2.2 (rAb, f1) (rAf, f3)	–	1.7 (r9b, f1)
Se	1.05 (r9d, f1)	2.4 (rAf, f3)		1.9 (r9b, f1)
Te	1.2 (r9d, f2)	2.5 (rAf, f3)	0.7 (rAg, f1)	2.1 (r9b, f1)

Appendix 4

The first-order collapse in SmS and (Sm/Gd) S bars one, obviously, from observing the full course of events through configuration crossover. It is for example not clear whether any rather sudden development of metallic behaviour, as from a Mott-Anderson transition, is involved here in promoting the first-order structural collapse and crossover. Results on SmSe and SmTe under pressure suggest that this is probably not so. The present resistivity data (*5*) (Fig. 8) for SmSe and SmTe under pressure shows no especially rapid move anywhere in the pressure region for which the volume is decreasing most steeply; i.e. at 6 – 1/2 kB for SmS, at ~ 35 kB for SmSe, and ~ 45 kB for SmTe (see Fig. 9)[22]. By these pressures ρ is down to 10^{-3} Ω – cm, a quite low value. Effects from configuration crossover in each case first appeared when $\rho \approx 10^{-2}$ Ω–cm (at which point the $f \rightarrow d_f$ gap is ~ 0.05 eV or 2 kT). If we take 1 kB as "onset" pressure in SmS, the corresponding values in SmSe and SmTe are 27 kB and 40 kB respectively.

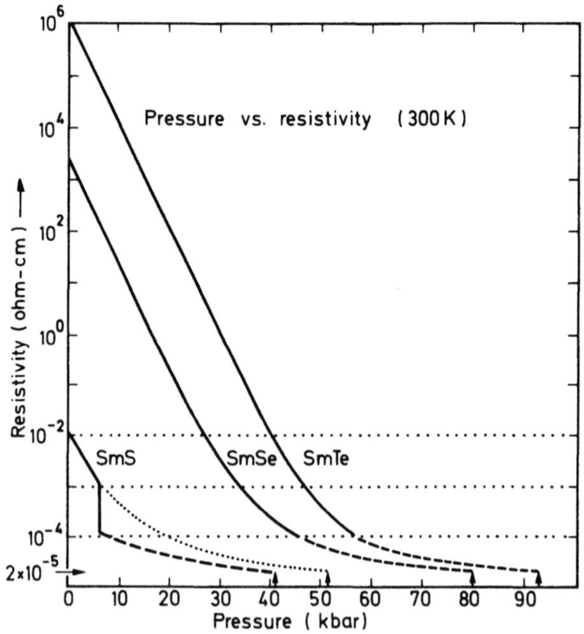

Fig. 8. Resistivity of samarium monochalcogenides versus pressure. The data from Ref. (*5*) has been extrapolated to $\rho = 2 \times 10^{-5}$ Ω–cm, which is taken as appropriate to full crossover. The behaviour of SmS that would occur, if no first order transition occurred, is indicated here, and again in Fig. 9.

[22]) In the series $SmS_{1-x}Se_x$ (*46*), the first-order transition detectable to $x = 0.8$, may be extrapolated on to 1.0. The value of 40 kB so adduced for SmSe is actually well *beyond* the point of steepest descent in Fig. 9. It could be that a discrepency in pressure calibration exists between (Refs. *5* and *46*), which should be checked out.

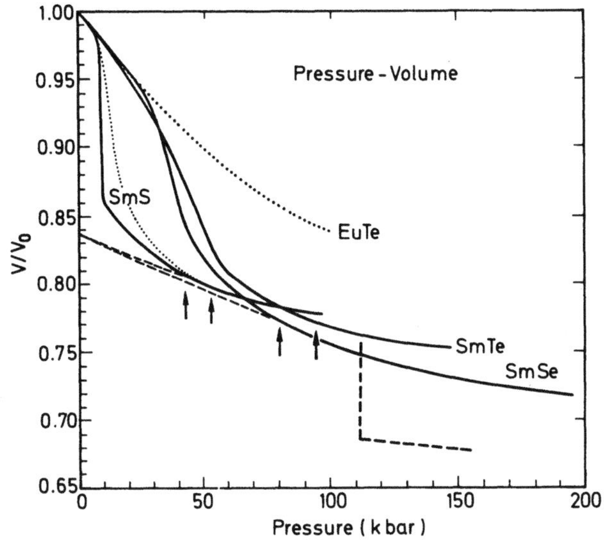

Fig. 9. Volume-pressure relationship in the samarium monochalcogenides and in EuTe (from Ref. *5a*). Extrapolation of the $f^5 d^1$ behaviour back to room pressure is given, starting from the indicated crossover completion points, these being the same as those deduced in Fig. 4.

Crossover is signaled as complete when ρ reaches a value much as for d^1 YS or LaS; say 2×10^{-5} Ω–cm at 300 K (*32b*). Extrapolation of the present data places completion in SmTe at 93 kB, in SmSe at 80 kB and in SmS at 41 kB. The first-order transition in SmS is seen to be equivalent to a pressure increment of 11 kB, ρ at 7 kB being what it would otherwise have been at 18 kB; completion of crossover is advanced from 52 kB to 41 kB.

The above completion pressures transpose satisfactorily from Fig. 4 to Fig. 5, in that extrapolation can then be made back, from pressures beyond these, to a fully collapsed volume ratio of 0.835 (equivalent to a lattice parameter for SmS of 5.66 Å in the fully collapsed $f^5 d^1$ condition, as was used in Fig. 2a. In both Figs. 4 and 5 we include the projected behaviour of SmS, as freed from its first-order transition. Also included in Fig. 5 is the data for EuTe, which (unlike EuS) has about the same initial compressibility as the expanded phase samarium materials. This, and the collapsed phase extrapolation above, provide reference lines against which to chart the course of crossover. The sigmoid character displayed here by SmSe is much as for high-spin low-spin crossover in 3*d* materials having a second-order transition. Using a Vegard's Law approach, as in § 2.1 for the SmS doped systems, leads to conversion percentages for SmS both before and after the first-order transition, that are much as in Fig. 2b.

The entire crossover process is found to be completed in the samarium materials at room temperature over a pressure span of close to 50 kB. This is again much as for the high-spin → low-spin transition of Fe^{2+} in the dilute pyrites solid solution MnS_2: Fe (*47*).

The resistivity of $Sm_{0.9}Y_{0.1}S$, a B-phase material under room conditions, is smaller by almost two orders of magnitude than for pure SmS, and, unlike SmS, it has a positive temperature coefficient (*32b*). It is however still close to one order greater than for the simple undisturbed d^1 metal YS.

References

1. For recent review see *C.M. Varma:* Rev. Mod. Phys., *48*, 218 (1976).
2. *Wilson, J.A.:* Adv. in Phys. *21*, 89 (1972); and 101–116 (1973) in the conf. report: Phase Transitions, 1973 (ed. *Cross, L.E.*) Pergamon.
3. *Hulliger, F.:* Helv. Phys. Acta, *41*, 945 (1968).
4. *Jørgensen, C.K.:* (a) Oxidation Number and Oxidation States. Berlin, Heidelberg, New York: Springer 1969; (b) Modern Aspects of Ligand Field Theory. North Holland-Elsevier 1971.
5. (a) *Jayaraman, A., Dernier, P.D., Longinotti, L.D.:* Phys. Rev. *B11*, 2783 (1975) and High Temps, High Pressures, 7, 1 (1975); (b) *Jayaraman, A., Singh, A.K., Chatterjee, A., Usha Devi, S.:* Phys. Rev. *B9*, 2513 (1974).
6. *Mattheiss, L.F.:* Phys. Rev. *B5*, 290, 306 (1972).
7. *Dumas, J., Schlenker, C.:* Phys. Stat. Sol. (a) *22*, 89 (1974).
8. *Mattheiss, L.F.:* (a) Phys. Rev., *B8*, 3719 (1973); (b) Phys. Rev. *B6*, 4718 (1972).
9. (a) *Frieser, M.J., Holtzberg, F., Methfessel, S., Pettit, G.D., Shafer, M.W., Suits, J.C.:* Helv. Phys. Acta., *41*, 832 (1968); (b) *Narayanamurti, V., Jayaraman, A., Bucher, E.:* Phys. Rev. *B9*, 2521 (1974); (c) *Guntherodt, G., Holtzberg, F.:* A.I.P. Conf. Rep. *24*, 36 (1975) (20th Mag. Conf., San Francisco, 1974); (d) *Vedam, K., Kirk, J.L.:* Proc. 11th R.E. Res. Conf., 840 (1973) (Michigan; USAEC); (e) *Kaldis, E., Wachter, P.:* Sol. St. Comm. *11*, 907 (1972).
10. *Wertheim, G.K., Mattheiss, L.F., Campagna, M., Pearsall, T.P.:* Phys. Rev. Lett., *32*, 997 (1974).
11. (a) *Konig, E., Ritter, G.:* Mossbauer Effect and Methodology 9, 34 (1974) in Plenum Press series, edited by *I.J. Gruverman, C.W. Seidel and D.K. Dieterly;* (b) *Richards, R., Johnson, C.E., Hill, H.A.O.:* J. Chem. Phys., *48*, 5231 (1968); (c) *Eibschutz, M., Di Salvo, F.J.:* Phys. Rev. Lett., *36*, 104 (1976).
12. (a) *Guntherodt, G., Wachter, P.:* A.I.P. Conf. Proc. *18*, 1034 (1974) (19th Mag. Conf., Boston, 1973); (b) *Hulliger, F., Hull, G.W.:* Sol. Stat. Comm. *8*, 1379 (1970).
13. *Sallach, R.A., Corbett, J.D.:* Inorg. Chem. *3*, 993 (1964).
14. *Mee, J.E., Corbett, J.D.:* Inorg. Chem., *4*, 88 (1964).
15. *Corbett, J.D., Sallach, R.A., Lokken, D.A.:* Adv. in Chem. (ACS ser) *71*, 56 (1967).
16. *Caro, P.E., Corbett, J.D.:* (Ref. 7), J. Less Common Metals, *18*, 1 (1969).
17. *Barnighausen, H., Patow, H., Beck, H.P.:* Z. Anorg. Allg. Chem., *403*, 45 (1974).
18. *Campagna, M., Bucher, E., Wertheim, G.K., Buchanan, D.N.E., Longinotti, L.D.:* Phys. Rev. Lett., *32*, 885 (1974).
19. *Tromme, M.:* C.R.Ac. Sci. (Paris), C *273*, 849 (1971).
20. *Longinov, G.M., Starovoitov, A.T., Golubkov, A.V.:* Fiz. Tv. Tela. *11*, 3637 (1969) (Sov. Phys. Sol. St. *11*, 3053 (1970).
21. *Francheschi, E., Olcesse, G.L.:* Phys. Rev. Lett., *22*, 1299 (1969).
22. *McTaggart, F.K.:* Australian J. Chem., *11*, 471 (1958).
23. *Scaife, D.E., Wylie, A.W.:* J. Chem. Soc. A, 5450 (1964).
24. *Suski, W.:* J. Sol. Stat. Chem. 7, 385 (1973).
25. *Yim, W.M., Stofko, E.J., Smith, R.T.:* J. Appl. Phys. *43*, 254 (1972).
26. *Jayaraman, A., Bucher, E., Lowe, W., Longinotti, L.D.:* Phys. Rev. Lett. *36*, 366 (1976).
27. *Brown, D.:* Halides of the Lanthanides and Actinides, Fig. 4.3 Wiley-Interscience, 1968.
28. *Hulliger, F.:* Structure and Bonding, *4*, 83 (1968), Fig. 24. Berlin, Heidelberg, New York: Springer 1968.
29. *Campagna, M., Bucher, E., Wertheim, G.K., Buchanan, D.N.E.; Longinotti, L.D.:* Proc. 11th R.E. Res. Conf., 810 (1973) (Michigan; USAEC).
30. (a) *Slater, J.C.:* Adv. in Quant. Chem. *6*, 1 (1972), § 8, (Academic Press) Ed. *Lowdin, P.O.;* (b) *Goscinski, O.:* Int. J. Quant. Chem. Symp. *9*, 221 (1975).
31. (a) *Eastman, D.E., Holtzberg, F., Freeouf, J., Erbudak, M.:* A.I.P. Conf. Proc., *18*, 1030 (1974) (19th Mag. Conf. Boston, 1973); (b) *Campagna, M., Bucher, E., Wertheim, G.K.,*

Longinotti, L.D.: Phys. Rev. Lett. *33*, 165 (1974); (c) *Freeouf, J.L., Eastman, D.E., Grobman, W.D., Holtzberg, F., Torrance, J.B.:* Phys. Rev. Lett., *33*, 161 (1974).
32. (a) *Holtzberg, F.:* A.I.P. Conf. Proc., *18*, 478 (Boston Mag. Conf. 1973); (b) *Penney, T., Holtzberg, F.:* Phys. Rev. Lett., *34*, 322 (1975); (c) *Tao, L.J., Holtzberg, F.:* Phys. Rev. *B11*, 3842 (1975).
33. *Campagna, M., Bucher, E., Wertheim, G.K., Longinotti, L.D.:* A.I.P. Conf. Proc. *24*, 39 (1974).
34. *Pearson, W.B.:* The Crystal Chemistry and Physics of Metals and Alloys, p 174. Wiley-Interscience, 1972.
35. *Coey, J.M.D., Ghatak, S.K., Holtzberg, F.:* A.I.P. Conference Proc. *24*, 38 (San Francisco Mag. Conf. 1974).
36. *Mott, N.F.:* Metal-Insulator Transitions, p 222. Taylor and Francis/Barnes and Noble, 1974.
37. *Kirk, J.L., Vedam, K., Narayanamurti, V., Jayaraman, A., Bucher, E.:* Phys. Rev. *B6*, 3023 (1972).
38. *Bader, S.D., Phillips, N.E., McWhan, D.B.:* Phys. Rev. *B7*, 4686 (1973).
39. *Barros, H.L., Chandrashekar, G.V., Chi, T.C., Honig, J.M., Sladek, R.J.:* Phys. Rev. *B7*, 5147 (1973).
40. *McWhan, D.B., Remeika, J.P., Rice, T.M., Brinkman, W.F., Maita, J.P., Menth, A.:* Phys. Rev. Lett. *27*, 941 (1971).
41. *Nugent, L.J., Van der Sluis, K.L.:* J. Chem. Phys., *59*, 3440 (1973).
42. *Di Salvo, F.J., Wilson, J.A., Waszczak, J.V.:* Phys. Rev. Lett., *36*, 885 (1976).
43. *Nickerson, J.C., White, R.M., Lee, K.N., Bachman, R., Geballe, T.H., Hull, Jr., G.W.:* Phys. Rev. *3*, 2030 (1971).
44. (a) *Holtzberg, F.:* A.I.P. Conf. Proc. *18*, 478 (Boston Mag. Conf. '73); (b) *Pollak, R.A., Holtzberg, F., Freeouf, J.L., Eastman, D.E.:* Phys. Rev. Lett. *33*, 820 (1974).
45. *Davis, H.M., Bransky, I., Tallan, N.M.:* J. Less Common Met. *22*, 193 (1970).
46. *Bucher, E., Maines, R.G.:* Sol. St. Comm. *11*, 1441 (1972).
47. *Bargeron, C.B., Avinor, M., Drickamer, H.G.:* Inorg. Chem. *10*, 1338 (1971).

References for A3

Aa. *Saum, G.A., Hensley, E.B.:* Phys. Rev. *113*, 1019 (1959).
Ab. *Scouler, W.J., Feinleib, J., Dimmock, J.O., Pidgeon, C.R.:* Sol. St. Comm., 7, 1685 (1969).
Ac. *Suryanarayanan, R., Paparoditis, C., Ferre, J., Briatt, B.:* Sol. St. Comm. *8*, 1853 (1970).
Ad. *Suryanarayanan, R., Ferre, J., Briatt, B.:* Phys. Rev. *B9*, 554 (1974).
Ae. *Beckenbaugh, W., Evers, J., Guntherodt, G., Kaldis, E., Wachter, P.:* J. Phys. Chem. Sol., *36*, 239 (1975).
Af. *Guntherodt, G.:* Phys. Kond. Mat. *18*, 37 (1974).
Ag. *Suryanarayanan, R., Guntherodt, G., Freeouf, J.L., Holtzberg, F.:* Phys. Rev. *B12*, 4215 (1975).
Ah. *Batlogg, B., Kaldis, E., Schlegel, A., Wachter, P.:* Phys. Lett. *56A*, 122 (1976).

The above work was prepared as a Bell Technical

Memorandum, August 1976.

I would like here to direct attention to the Proceedings of the International Conference on Valence Instabilities held at Rochester, New York, in November 1976, shortly to be released by Pergamon Press. A further illustrative appendix to the present text appears in those proceedings.

The Potential Energy Surfaces of Polyatomic Molecules

J.N. Murrell

School of Molecular Sciences, University of Sussex, Falmer, Brighton BN1 9QJ.

Table of Contents

Preface

The object of this review is to give a broad perspective of potential energy surfaces for polyatomic molecules without any attempt at a comprehensive coverage of any one area. Several of the sections are worthy of being treated as monographs in their own right. The review is based upon a series of lectures given at a Summer School on Physical Organic Chemistry in October 1974 organised by the Accademica Nationale dei Lincei. Only section 6 has been substantially modified in the light of recent developments.

1. Introduction

The molecular potential energy surface is one of the most important concepts of physical chemistry. It is at the foundations of spectroscopy, of chemical kinetics and of the study of the bulk properties of matter. It is a concept on which both qualitative and quantitative interpretations of molecular properties can be based. So firmly is it placed in the theoretical interpretation of chemistry that there is a tendency to raise it above the level of a concept by ascribing it some physical reality.

Although the idea of forces between atoms was current in chemical thinking before the development of modern quantum mechanics, a rigorous definition of a molecular potential energy surface was first given by *Born* and *Oppenheimer* (*1*) through their analysis of the separation of the total molecular wave function into electronic and nuclear parts. In the same year the first approximate solutions of Schrodinger's equation for the electronic states of a molecule by *Burrau* (*2*), for H_2^+, and by *Heitler* and *London* (*3*), for H_2, gave the first approximate potential energy curves for diatomic molecules.

London (*4*) was quick to see the implications of these calculations for the theory of chemical reactions. He suggested that many chemical reactions were electronically adiabatic. That is, there is a smooth change of the electronic state as the nuclei move from their reactant to their product configuration, so that the reaction takes place on a single potential energy surface. He considered moreover that future calculations using quantum mechanics would lead to perfect agreement between calculated and experimental activation energies; this showed greater foresight than he probably appreciated at the time for the day of realisation is still far ahead for all except the simplest molecular systems.

In the past fifty years we have gathered detailed knowledge about potential energy surfaces from many branches of chemistry and physics: in particular from theoretical valence calculations, from spectroscopy, from kinetics, particularly the "non-

Boltzmann" molecular beam experiments, and from the bulk properties of matter. In addition, the theoretical concept of a surface has been put on firmer grounds so that its role as a basis for a quantitative analysis of experimental results is now clearer.

2. Definition of a Potential Energy Surface

The molecular potential energy is an energy calculated for static nuclei as a function of the positions of the nuclei. It is called *potential* energy because it is the potential energy in the dynamical equations of nuclear motion.

The position of nuclei in space are defined by $3N$ independent coordinates. If three coordinates are used to specify the centre of mass, then the remaining $3N-3$ coordinates may be referred to this as origin. In the absence of external fields the energy and relative motion of the nuclei are independent of the centre of mass and so these three coordinates can be ignored in any dynamical calculation.

A further three coordinates specify the orientation of a solid body to axes fixed in space. The potential energy may therefore be reduced to a function in $3N-6$ dimensions (V) (except for diatomic molecules for which only two dimensions are required to specify the orientation of the nuclei in space). However, nuclear motion is a $3N-3$ dimensional problem, and if one solves separately the rotational motion by introducing the angular momentum constants of motion then the effective potential for the remaining $3N-6$ degrees of freedom will consist of V plus terms arising from centrifugal forces. For example, for a diatomic molecule in a rotational state of quantum number J the effective vibrational potential is (ignoring electron orbital or spin contributions to J)

$$U(R) = \mathrm{V}(R) + \frac{\hbar^2}{2\mu R^2} J(J+1) . \tag{1}$$

where μ is the reduced mass of the diatomic molecule.

The separation of the dynamical motion of electrons and nuclei is possible because of the large difference in masses between the two. The separation is referred to as the Born-Oppenheimer approximation or principle (*1*). In order to understand its limitations we must express it in a precise relationship to the quantum mechanical equations of motion.

The non-relativistic Hamiltonian for n electrons (i) and N nuclei (μ) in terms of space fixed axes, has the form

$$H = \sum_i T(i) + \sum_\mu \mathrm{T}(\mu) + V(i, \mu) . \tag{2}$$

where the kinetic energy terms are separated for each particle and where the potential energy V depends only on interparticle distances. To solve the quantum mechanical equations we make the Born-Oppenheimer approximation and we also separate 3 coordinates for the centre of mass of the molecule. The precise form of the Hamiltonian depends on the order in which we do these operations. If, for example, we first make the Born-Oppenheimer approximation, then the electron coordinates are referred to some geometrical centre of the nuclei, and the centre of mass is approximated by the centre of mass of the nuclei. There is then a small term in the Hamiltonian (the so-called mass polarization term) which corrects for the fact that the true centre of mass depends on the electron mass as well as the nuclear masses. Alternatively, one can first define a true centre of mass and measure the electron coordinates relative to this. In this way the mass polarization term may be removed from the Hamiltonian. However, the resulting electronic wave functions which are obtained are then dependent upon the nuclear masses as well as the nuclear charges and such wave functions are an inconvenient basis from which to investigate nonadiabatic processes.

It is necessary to worry about the relative merits of these two approaches if one is concerned with non-adiabatic processes (Some references on this point are *5–9*), but, as our primary concern is with the adiabatic potential energy surfaces, we can ignore the difference, which effectively means that we assume the mass of the electron to be zero.

The description we give of the Born-Oppenheimer approximation is roughly that followed by *Born* and *Huang* (*10*) and others. As a basis to the eigenfunctions of the full molecular Hamiltonian (2) we take a set of functions[1])

$$\psi_n(r_i, r_\mu)\, \theta_n(r_\mu)\,. \tag{3}$$

$\psi_n(r_i, r_\mu)$ is an eigenfunction of the electronic Hamiltonian, which is the total Hamiltonian without the nuclear kinetic energy terms.

$$H = H_e(r_i, r_\mu) + \sum_\mu T(\mu)$$

$$H_e(r_i, r_\mu)\, \psi_n(r_i, r_\mu) = E_n(r_\mu)\, \psi_n(r_i, r_\mu)\,. \tag{4}$$

[1]) The basis states

$$\psi_n(r_i,\ r_\mu^0)\, \vartheta_{\hat{n}}(r_\mu)$$

in which the electronic functions are referred to a fixed and convenient nuclear configuration (r_μ^0) gives a zeroth order (i.e. a single term) approximation to the exact wavefunction which is often referred to as the crude Born-Oppenheimer approximation. It was in fact the basis used in reference 1. *Englman* (*10a*) emphasized that this basis has some advantages over the "adiabatic" basis (3) in situations where the Born-Oppenheimer approximation breaks down, although the basis is very slowly convergent for large displacements from (r_μ^0).

ψ_n depends parametrically on the positions of the nuclei r_μ.
We then seek a solution of the full wave equation of the form:

$$(H - E) \sum_n \theta_n(r_i, r_\mu)\, \theta_n(r_\mu) = 0 . \tag{5}$$

Multiplying this equation by each electronic function in turn and integrating over the electronic coordinates leads to a set of coupled differential equations in the nuclear coordinates as follows

$$\sum_n (H_{n'n}(r_\mu) - E\, \delta_{n'n})\, \theta_n(r_\mu) = 0 . \tag{6}$$

which from (4) and (5) have the form

$$(E_{n'} - E)\, \theta_{n'}(r_\mu) + \sum_n \sum_\mu \int d\tau_i\, \psi_{n'}(r_i, r_\mu)\, T(\mu)\, \psi_n(r_i, r_\mu)\, \theta_n(r_\mu) = 0 . \tag{7}$$

$T(\mu)$ being a differential operator acts on ψ_n and θ_n. As it is a second-order differential we introduced $T^{1/2}$ as

$$T_\mu^{1/2} = -\frac{1}{(2M_\mu)^{\frac{1}{2}}} \nabla_\mu . \tag{8}$$

It is convenient to put the terms depending on the rate of change of ψ_n with r_μ on the right hand side of (7) in order that, when ψ_n is a slowly varying function of r_μ, they may be dropped. The coupled equations therefore have the form

$$\left[\sum_\mu T(\mu) + (E_{n'}(r_\mu) - E)\right] \theta_{n'}(r_\mu)$$
$$= - \sum_n \sum_\mu [< \psi_{n'} | T(\mu) | \psi_n > + 2 < \psi_{n'} | T(\mu)^{1/2} | \psi_n > \cdot T(\mu)^{1/2}]\, \theta_n(r_\mu) . \tag{9}$$

The bracket notation $<>$ indicates integration over the space and spin coordinates of the electrons.

If the right hand side of (9) is dropped, then the equations are uncoupled and each equation defines a set of nuclear wave functions $\theta_{n,k}(r_\mu)$ as eigenfunctions of the operator

$$\left[\sum_\mu T(\mu) + E_n(r_\mu)\right] \tag{10}$$

in which the electronic energy $E_n(r_\mu)$ behaves as the potential energy for the nuclear motion. It is this function, subject to separation of the translational and rotational coordinate, that we shall understand as the potential energy surface in the Born-Oppenheimer approximation. It is also possible to uncouple Eq. (9) by dropping all terms on the RHS except $n = n'$. In this case we still have an adiabatic approximation in which one electronic state defines a surface, but the surface now has contributions from the electronic matrix elements of T and $T^{1/2}$. These terms are not generally considered as having great physical significance – they are not as important as the coupling terms ($n \neq n'$) for example which are responsible for non-adiabatic processes[2]).

In circumstances where the coupling terms in equation (9) are large it is often possible to define another set of electronic states $\phi_n(r_i, r_\mu)$, say, for which they are smaller. This will be at the expense of not having the basis ϕ_n as eigenfunctions of H_e, so that there will now be "electronic" coupling terms between the basis functions. Nevertheless, the sum of these electronic and nuclear coupling terms may be smaller than the nuclear coupling terms in the Born-Oppenheimer approximation. Potential energy surfaces $E_n(r_\mu)$ defined in this way are surfaces called diabatic potentials[3]).

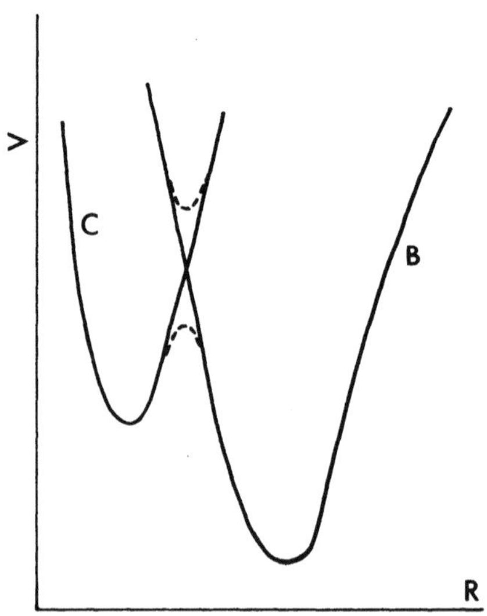

Fig. 1. Diabatic (full lines) and adiabatic (dashed lines) potentials for the B and C, $^2\Pi$, states of NO.

---

[2]) *Kolos* and *Wolniewicz* (*11*) calculate from accurate wave functions of the ground state of H_2 that the diagonal nuclear kinetic energy term is 5 cm^{-1} at the equilibrium distance and – 38 cm^{-1} at a nuclear separation of 0.3 Å.

[3]) They have been of considerable importance in the discussion of electron scattering from atoms and molecules (*12–14*).

Diabatic surfaces connect electronic states having similar electron configurations. In other words the one-electron functions from which the ϕ_n are constructed change little as the nuclear coordinate of interest is changed.

As an example we can take the excited states of NO. It has been shown that there are two excited states of the same symmetry ($^2\Pi$) whose vibrational levels are best interpreted on the basis of diabatic curves which cross as in Fig. 1 (*15–17*). One of these states (*B*) arises from the electron excitation to an antibonding valence molecular orbital and the other (*C*) from excitation to a Rydberg orbital. The Born-Oppenheimer adiabatic curves cannot cross (by virtue of the non-crossing rule which is to be discussed in a later section) and must fullow the dashed curves shown in the figure. It is concluded that the coupling of the electronic vibrational basis is much smaller in the diabatic than in the adiabatic basis, by virtue of the observation that the diabatic vibrational spacing is similar to that of the observed spacing, and this is because there is little mixing between Rydberg and valence orbital configurations. However, it must be emphasized that although diabatic potentials are a good basis for understanding some molecule properties there is, unlike for the adiabatic potentials, in most cases no rigorous definition of such potentials, or their defining Hamiltonian.

Finally, we note that there are some situations, of which the dynamical Jahn-Teller and Renner-Teller effects are the classic examples, where the nuclear coupling terms are important in any electronic basis (*18*). Adiabatic potential surfaces may still be defined but the resulting molecular eigenstates cannot be identified with any one surface.

3. The Choice of Coordinates

If we examine a potential energy surface there are several features which play an important role in the interpretation of kinetic processes. These are minima (stable configurations of all the atoms), valleys (separate stable groups of atoms which we identify as reactants and products) and saddle points (transition states). However, before we give a more formal definition of these features we have to consider the coordinate system that is used.

We can choose as $3N-6$ coordinates any set which are independent of the absolute position and orientation of the nuclei in space. In practice the most convenient coordinates will consist of internuclear distances or bond angles, but there is no unique choice and one must consider the convenience for the particular problem in hand. We will illustrate this in two cases.

(i) Three nuclei (ABC). We require three internal coordinates and these could be the three internuclear distances (r_{AB}, r_{BC}, r_{CA}) or two distances and the angle between these ($r_{AB}, r_{BC}, A\hat{B}C$) or even one distance and two angles. If we are interested in a non-reactive atom-diatom collision or in a chemical reaction

$$AB + C \rightarrow A + BC$$

then either of the first two suggestions is convenient. However, if we wish to include also the possibility of reaction to a third channel $AC + B$, then it is better to use the three distances as each of the three products is being treated on the same basis.

(ii) A five atom molecule. The point to be made here is that the surface is 9-dimensional but there are 10 internuclear distances, so that one of these is a determined function of the other nine. The choice of coordinates now depends upon the problem in hand, but no set is ideal for all possibilities.

In considering the dynamics of potential energy surfaces an important feature is the fact that the internal coordinates are not in general an orthogonal set. By this, one means that the Hamiltonian has cross terms in its kinetic energy part. For example, the kinetic energy of the linear triatomic (nonrotating) system.

$$A \overset{r_1}{—} A \overset{r_2}{—} A$$

expressed in the internuclear coordinates r_1 and r_2 is (in classical mechanics)

$$T = \frac{m}{3}(\dot{r}_1^2 + \dot{r}_1^2 + \dot{r}_1 \dot{r}_2) . \tag{11}$$

If one plots classical trajectories on a potential energy surface with r_1 and r_2 as orthogonal coordinates then the cross terms in the kinetic energy lead to an apparently strange motion, the trajectory not being one expected for a particle moving freely on the surface (for which there would be no cross term). This in illustrated by Fig. 2a.

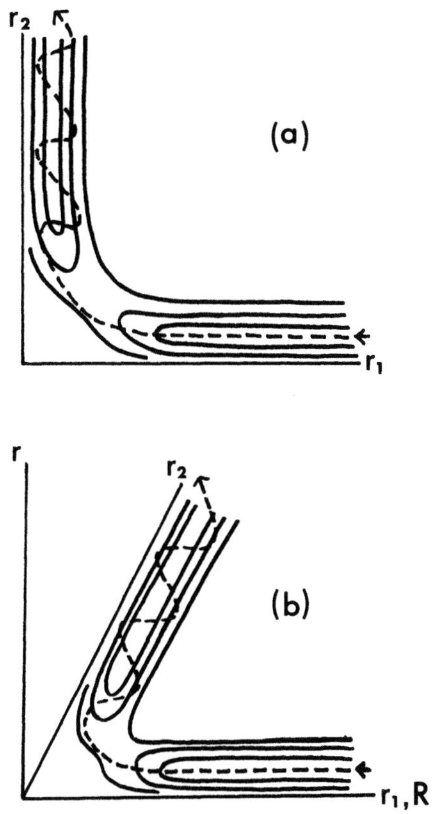

Fig. 2. Classical trajectory for a colinear $A + A_2$ collision plotted in non-orthogonal (*a*) and orthogonal (*b*) coordinates.

However, it is always possible to define new coordinates, which are linear combinations of the internal coordinates, for which there are no cross terms in the kinetic energy, and there are an infinite number of such sets. To illustrate this for the example given, we can take as new coordinates, the distance between a terminal atom and the centre of mass of the other two, namely

$$R = r_1 + \frac{r_2}{2} \tag{12}$$

and a scaled distance between the other two atoms,

$$r = \frac{\sqrt{3}}{2} r_2 . \tag{13}$$

The kinetic energy then occurs in diagonal form

$$T = \frac{m}{3}(\dot{R}^2 + \dot{r}^2)\,. \tag{14}$$

It can be seen from Fig. 2 that by adopting this transformation we have effectively chosen to construct the potential energy surface in a system in which the internal coordinates are skewed. Calculated trajectories now correspond to those of a freely moving particle, of mass 2 m/3, on the surface. However, although such coordinates simplify the kinetic energy term, they may complicate any algebraic expression we have for the potential energy. In addition the boundary conditions to be imposed on eigenfunctions of the problem are more complicated.

In the quantum mechanical treatment of reactive processes the choice of coordinates plays an important role in the solution of the problem. This is because there is one direction on the surface, which chemists would generally associate with the reaction coordinate, along which the wave function is oscillating rapidly (depending on the translational motion of the incoming and outgoing particles) and other directions (corresponding to vibrations) in which the oscillations are slow. It is useful if a set of coordinates is chosen which roughly separates these two types of motion.

For the linear reactive system

$$A + BC \rightarrow AB + C$$

the reaction coordinate starts out by being paralled to r_{AB} and ends by being parallel to r_{BC}. Over its whole range it must be represented by a curvilinear transformation of the internal coordinates. Such coordinates were first defined in a rigorous manner by *Marcus* (*19*), and they have been generalized by *Light* (*20*). The detailed form of the kinetic energy operator is reasonably simple only for the case of the linear triatomic molecule, and in view of the importance of this in the literature of reactive scattering, an outline of the mathematics is appropriate at this point.

Let any smooth curve C be defined that leads asymptotically from the reaction to the product valley with curvature of only one sign. Coordinates may be defined by the perpendicular distance of a point from this curve (x) and the distance of the projection along the line (s). In these coordinates the classical kinetic energy for internal motion of the particles is

$$T = \tfrac{1}{2}\mu\{\eta^2\,\dot{s}^2 + \dot{x}^2\} \tag{15}$$

where μ is the mass of the representative point on the surface (2 m/3 for three equal masses) and where η is related to the curvature $\kappa(s)$ of the curve C by

$$\eta = 1 + \kappa x\,. \tag{16}$$

The corresponding quantum mechanical operator is

$$-\frac{\hbar^2}{2\mu}\left\{\frac{1}{\eta}\frac{\partial}{\partial s}\left(\frac{1}{\eta}\frac{\partial}{\partial s}\right)+\frac{1}{\eta}\frac{\partial}{\partial x}\left(x\frac{\partial}{\partial x}\right)\right\} \tag{17}$$

which we note is not separable because η depends on both x and s. However, if the curvature κ of the curve C is nowhere very large then $\eta \simeq 1$ and the kinetic energy is approximately separable. There will however be, in general, coupling between the x and s motion through the potential energy terms.

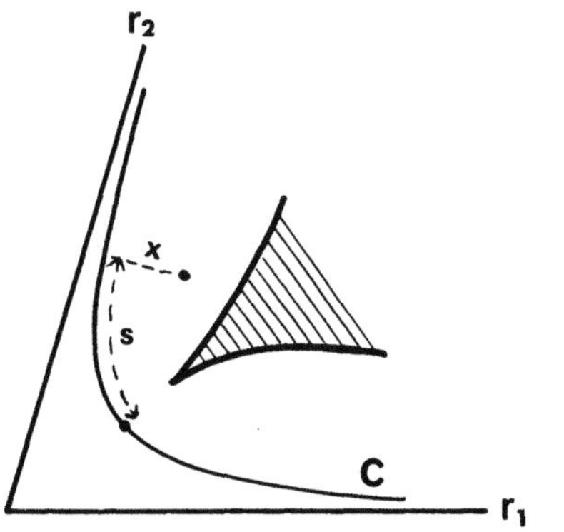

Fig. 3. Curvilinear coordinate system introduced by *Marcus* (*19*) and *Light* (*20*) showing the hatched region in which the coordinates x and s are not uniquely defined.

Marcus (*19*) takes the curve C to lie along the bottom of the reaction valley but *Light* (*20*) prefers to define it analytically in terms of the internal coordinates so that explicit expressions may be given for η. One problem with such curvilinear coordinates is that they are not single valued in certain regions of space. Fig. 3 shows the locus of centres of curvature of the curve C. Inside the hatched region each point lies on two or more lines which are perpendicular to C, and hence in this region x and s are not uniquely defined. In practice this does not matter providing that the potential in the hatched region is so high that the trajectories never penetrate to these parts. As a general rule one should not adopt a curve C that has large curvature at any point.

4. Characteristics of Potential Energy Surfaces

A stable nuclear configuration on a potential energy surface is associated with a point for which there is zero slope in any direction and for which there is no direction in which the curvature is negative or zero. Such points are uniquely defined in any system of internal coordinates but we shall see that some other characteristic features of a surface are dependent on the choice of coordinate.

Reactants or products in a bimolecular system are represented as asymptotic regions of the surface (one or more internal coordinates R_i becoming infinite) at which the potential is independent of these coordinates R_i. For example, the bimolecular reaction $AB + CD$, has a reactant "valley" which is "flat" in four dimensions: V is independent of $R_{AC}, R_{AD}, R_{BD}, R_{BD}$ when these coordinates become infinite. If lines are drawn on the surface which pass from reactant to product "valleys" (or between stable reactant and product configurations) then for each chemical process there are one or more points which represent maxima on minimum activation energy paths. These maxima are transition states for the reaction. However, as we shall see, it is more difficult (at least in many dimensions) to establish that a point represents a transition state than that it represents the configuration of a stable species.

Transition states are generalized saddle points on many-dimensional surfaces. The slope of the potential is zero for all directions and only one of the $3N-6$ principal curvatures is negative. As it is an important fact that only one curvature is negative we must define what is meant by principal curvatures. At any point on the surface we can establish a matrix of second derivatives of the potential (force constants)

$$F_{ij} = \partial^2 V / \partial R_i \partial R_j \,. \tag{18}$$

It is always possible to transform to an orthogonal set of coordinates in terms of which the matrix F will be diagonal. If only one of the diagonal elements of this matrix is negative, and all first derivatives of V are zero, then this is a saddle point and a feasible transition state for some reaction. The only additional complication arises from the fact that the internal coordinates R_i are not in general orthogonal (as we have already seen) so that the transformation to an orthogonal set involves a knowledge of nuclear masses and geometry. The problem may be handled by the standard FG method of constructing normal modes of vibration (*21*), the only difference being that a saddle point is found to have one negative eigenvalue of the FG matrix. The normal coordinates Q_i which are the eigenvectors of FG are mass weighted distances such that the kinetic energy is $\sum_i \dot{Q}_i^2/2$.

The concept of a reaction coordinate has played a large part in qualitative discussions of chemical reactions and it has recently been used as a basis for symmetry rules. For this reason it is useful to attempt a rigorous definition of the term. This is done whilst recognizing that the actual path or trajectory that a chemical reaction takes on a potential energy surface (we use here the concepts of classical mechanics) depends upon the total energy and initial conditions for the trajectory.

At the transition state the reaction coordinate Q_r is defined to be in the direction of the one normal coordinate which has a negative eigenvalue. *Fukui* (*22*) suggested that the reaction coordinate be defined as the line of steepest descents from the transition state to reactant or product regions on the potential surface. However, this definition fails to acknowledge the non-orthogonality of the internal coordinates and it would appear more satisfactory to define the reaction coordinate with respect to normal coordinates Q for all points on its path so that the line of steepest descent will be that for which $(\partial V/\partial Q_r)$ is a maximum (correspondingly all other derivatives $(\partial V/\partial Q_s)$ $s \neq r$ will be zero).

It would be computationally convenient to have a recipe for plotting the reaction coordinate by starting in the reactant or product regions (whose configurations are generally known) rather than from the transition state (whose configuration is not known). This is unfortunately not always possible, the reason lying in the fact that the line of steepest ascent is not defined for a point at which grad $V = 0$. At the transition state, a point for which this is the case, the direction of the reaction coordinate is defined by the direction of negative curvature, but in a reactant or product valley the reaction coordinate may only be asymptotic to the direction of minimum curvature, i.e. it is asymptotic to the line for which grad V is a minimum. This situation is illustrated by the potential for $H_2 + H_2$ in rectangular configurations shown in Fig. 4. We shall have more to say about this potential when we discuss symmetry forbidden reactions.

Fig. 4. Potential energy surface for the ^1A_g state of rectangular H_4. The full line is the reaction coordinate and the dashed lines lead up blind valleys.

5. Methods of Finding Extrema on Potential Energy Surfaces

There are standard methods for finding the positions of minima (or maxima) on many-dimensional surfaces. If there is no foreknowledge of the approximate position of the minimum, which is rare in potential energy problems, then one has to start by a mapping technique or pattern search, the most efficient of which appears to be that known as the Simplex procedure (*23, 24*).

A simplex is a geometrical figure defined by $n + 1$ vertices in n dimensional space (e.g. in two dimensions this would be a triangle). The procedure is to calculate the energy at $n + 1$ points and then to replace the worst point P_{n+1} (the point of maximum energy) by another P^* which is the other side of the hyperplane defined by the remaining n points. If P is the centroid of these n points, then the new point is obtained by inverting P_n through $\bar{P}$; that is

$$(P^* - P) = (P - P_{n+1}) .$$

Providing the energy at P^* is not again the worst of the set, the procedure is simply repeated with the worst point until the required minimum is obtained. If however P^* is again the worst point then some other point on the line $P_{n+1} - P$ must be found which is not the worst point, and the process continued.

The simplex procedure has been tested on difficult surfaces with spiral valleys and found superior to the older established theories such as the method of steepest descents. If however the surface is well behaved or if by an intelligent guess one can get close to the minimum then the older methods can be efficient.

In the method of steepest descents one calculates the gradient at a point. The method of attack depends on whether this gradient may be calculated analytically or numerically (which requires calculations at $N + 1$ points for an N dimensional surface) and one moves along this direction until the lowest point is reached when a new gradient is calculated. When one is close to the minimum and the gradient is small it is necessary to have a method which is quadratically convergent, and to calculate the general quadratic function for N dimensions numerically requires $(N + 1)(N + 2)/2$ function evaluations.

It is not appropriate in this review to discuss the relative efficiency of minimization routines based on steepest descent procedures, but a few of the more important references are given. If the gradient can be evaluated analytically, and as *McIver* and *Komornicki* (*25*) have pointed out this is the case for many surfaces calculated by semi-empirical molecular orbital methods, then this is the best approach. The methods developed by *Davidson* (*26*), *Fletcher* and *Powell* (*27*), and *Murtagh* and *Sargent* (*28*) assume this to be the case. If the gradient has to be obtained numerically then the method developed by *Powell* (*29*) has found wide acceptance for its efficiency. *Garton* and *Sutcliffe* (*30*) have recently given a full account-of minimization methods with a wider brief than potential energy surfaces.

We have already emphasized that it is more difficult to establish that a point is a transition state than to establish that one is a position of equilibrium. It is therefore not surprising that techniques for finding such a point directly are not well established. The most common approach has been to start from a position of equilibrium or from a product or reactant valley, and to advance up the reaction coordinate, but we have already seen in the example of figure 4 that this will fail in some cases.

A technique used by Dewar and co-workers is to select any internal coordinate that changes during the reaction and to define this as the reaction coordinate. For selected values of this coordinate the energy is minimized with respect to all other internal coordinates. Thus for the ring opening of cyclopropane *Dewar* and *Kirschner* (*31*) took the $C\hat{C}C$ bond angle as reaction coordinate. This approach is likely to be successful only if the chosen coordinate is close to being the true reaction coordinate over the whole reaction. If it is not then one may find apparent blind alleys on the surface or discontinuous energy profiles. Dewar and Kirschner introduced the term chemical hysteresis for a reaction profile obtained by this method which depends upon the direction which one is advancing along the reaction coordinate. Fig. 5 shows a typical situation for the linear atom-diatom ($A + BC \rightarrow AB + C$) reaction. If the AB bond length is taken as reaction coordinate then the path calculated will be the dotted line where the tangents to the energy contours are parallel to r_{bc}. The line of steepest descent from the transition state however is the full line passing through the points of maximum curvature of the contours. The method of *Dewar* and *Kirschner* would seem to fail in too many cases for it to be adopted as a general procedure.

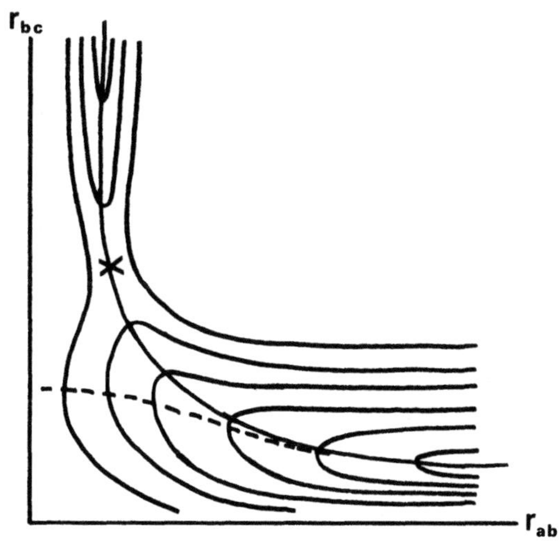

Fig. 5. Comparison of the reaction coordinate which passes through the transition state (x) and the dashed line obtained by minimizing the energy with respect to r_{bc} for each value of r_{ab}.

6. Influence of Symmetry

Symmetry has for many years played a vital role in the elucidation of molecular structure although apart from some special situations it was not thought to have a dominant influence on the structure or chemical properties of molecules. In recent years however it has played a large part in the interpretation of many organic reactions through the work of Woodward and Hoffman, and the concept of symmetry allowed or forbidden reaction is now an important part of mechanistic organic chemistry.

Historically the first application of symmetry to potential energy surfaces was to prove the so-called non-crossing rule. In its simplest form this may be stated as "potential energy curves for states of diatomic molecules of the same symmetry do not cross". We have already seen in section 2 that this should be qualified to apply to adiabatic curves, as in some situations it may be convenient to define diabatic curves which do cross.

The established methods of proving the non-crossing rule which are in the literature (*32, 33*) have recently been criticised by *Naqvi* and *Byers-Brown* (*34*), but their proof in turn has been criticized by *Longuet-Higgins* (*35*). All proofs attempt to show that crossing will only occur if conditions are satisfied for two functions of the internuclear distance R *which are assumed to be independent*. If we write this in the general form

$$f(R) = 0, \quad g(R) = 0 \tag{19}$$

then even if f and g are individually zero at discrete points in the range of R, it would only be by accident that they are zero simultaneously: the non-crossing rule cannot rule out accidents. However, if there is some special reason why one of these functions is zero for all values of R, and that can be so if the two states involved have different symmetries, then we may find a point where the other function is zero also and this is a crossing point.

The Teller proof (*33*) assumes that the eigenfunctions of the two states of concern (ψ_a and ψ_b say) may be written as a linear combination of two orthonormal basis functions σ_a and σ_b. The energies of the two states are then identical to the eigenvalues of the 2 × 2 Hamiltonian matrix

$$\begin{pmatrix} H_{aa} & H_{ab} \\ H_{ab} & H_{bb} \end{pmatrix} \tag{20}$$

which are given by the expression

$$2E = H_{aa} + H_{bb} + [(H_{aa} - H_{bb})^2 + 4H_{ab}^2]^{1/2} . \tag{21}$$

Both eigenvalues are therefore identical only if the term in brackets is zero and as this is a sum of two squared terms these must be separately zero. Thus our two conditions are

$$\begin{aligned} f(R) &\equiv (H_{aa} - H_{bb})(R) = 0 \\ g(R) &\equiv H_{ab}(R) = 0 \,. \end{aligned} \qquad (21a)$$

If σ_a and σ_b have different symmetries then H_{ab} is always zero and the curves therefore cross when $H_{aa} = H_{bb}$.

One might argue at this point that σ_a and σ_b may not be a complete basis for the expansion of the two eigenfunctions. Teller surmounted this by defining $\sigma_a, \sigma_b, \psi_r(r \neq a, b)$ to be a complete orthogonal set whence, as ψ_a and ψ_b cannot contain elements of the $\psi_r(r \neq a, b)$, σ_a and σ_b must be complete for ψ_a and ψ_b. However, the criticism of *Naqvi* and *Byers-Brown* (*34*) of the Teller proof lies in just this restriction. because with it σ_a and σ_b can equally well be written as some combination of ψ_a and ψ_b. It is then easy to show that if one of the conditions (21a) is satisfied the other must be satisfied also, which conflicts with our premise that the zeros of f and g are unconnected. The Teller proof therefore leaves us in the unusual position of having established that approximate potential energy curves of the same symmetry do not cross but leaves unanswered the question of whether the exact curves can cross.

Longuet-Higgins has extended the Teller proof by considering the eigenvalues of the symmetric matrix obtained from a general orthonormal basis of N states $\sigma_a, \sigma_b, \sigma_c, \ldots$. For such a matrix, H_{ij}, to have a doubly degenerate root E_0 the rank of the matrix $(H_{ij} - E_0 S_{ij})$ must be $N - 2$, that is, all its first minors must be zero (*36*). *Longuet-Higgins* then shows that for a symmetric matrix sufficient condition for this is that E_0 be a root of each of the equations.

$$\begin{vmatrix} H_{11} - E, H_{13}, \ldots \\ H_{31}, H_{33} - E, \ldots \\ \ldots \end{vmatrix} = 0$$

$$\begin{vmatrix} H_{12}, H_{13}, \ldots \\ H_{32}, H_{33} - E, \ldots \\ \ldots \end{vmatrix} = \begin{vmatrix} H_{21}, H_{23} \\ H_{31}, H_{33} - E, \ldots \\ \ldots \end{vmatrix} = 0$$

$$\begin{vmatrix} H_{22} - E, H_{23}, \ldots \\ H_{32}, H_{33} - E, \ldots \\ \ldots \end{vmatrix} = 0$$

These three conditions are clearly independent because each involves a matrix element that does not occur in the others. By eliminating E_0 between them we are left with two independent conditions that there should be a double root of any value.

For polyatomic molecules with $3N-6$ internal degrees of freedom it has been generally agreed (*33, 36a, 37*) that it would be possible to satisfy two conditions of the type established above and therefore that potential energy surfaces could cross even if they had the same symmetry. For example, a linear triatomic has two internal degrees of freedom and hence surfaces of the same symmetry could be expected to cross at a point. This supposition was called into question by *Naqvi* (*38*) who based his conclusion on an extension of the diatomic proof (which was based upon perturbation theory) given by himself and *Byers-Brown* (*34*). However, *Naqvi's* proof has been refuted by *Longuet-Higgins* (*35*) on the grounds that such intersections on polyatomic surfaces cannot in fact be treated by perturbation theory as their energies and wave functions are not analytic functions of the internal coordinates at the intersection points. This was earlier shown to be true for the conical intersections characteristic of the Jahn-Teller effect which we shall discuss shortly (*37*).

Longuet-Higgins (*35*) has shown that if the sign of an electronic wave function changes when passing round a closed loop in configuration space then there must be a point within the loop where states are degenerate. Moreover this condition is certainly satisfied by the states of some unsymmetrical triatomic molecules, and *Longuet-Higgins* establishes this for a molecule constructed from three atoms with ^2S ground states (e.g. HLiNa). In a D_{3h} configuration of H_3 the ground state is spatially degenerate (^2E). On lowering the symmetry to C_s this state splits into two $^2A'$ states so that one can say that there is a crossing, of the two surfaces in D_{3h} configurations which is allowed by symmetry. The molecule HLiNa would also have two low-energy $^2A'$ surfaces and as *Longuet-Higgins* has shown these are expected to cross along a line in three coordinate space. It is important to contrast the situation of a diatomic and polyatomic molecule. Lowering the symmetry of a diatomic (say from A_2 to AB) will convert some curve crossings to avoided crossings. Lowering the symmetry of a polyatomic (say from A_3 to ABC) will cause surface crossings to move to other parts of configuration space but will not necessarily convert them to avoided crossings.

The first theorem relevant to molecular structure based upon symmetry was by *Jahn* and *Teller* (*39*). They proved that the electronic wave function appropriate to a minimum on a potential energy surface could not have spatial degeneracy except for linear configurations of a molecule. Conversely, for a degenerate electronic state the potential energy surface must have a non-zero gradient at any configuration in which the degeneracy occurs.

To prove the Jahn-Teller theorem and to discuss related phenomena we consider the change in the electronic Hamiltonian of a molecule by making small distortions of the nuclei from some chosen origin. For convenience these distortions are represented by vectors in the space of the normal coordinates Q_i of the original structure. The change in the Hamiltonian can therefore be written as the Taylor expansion

$$H(Q_i) - H(0) = \sum_i \left(\frac{\partial H}{\partial Q_i}\right)_0 Q_i + \tfrac{1}{2} \sum_i \sum_j \left(\frac{\partial^2 H}{\partial Q_i \partial Q_j}\right)_0 Q_i Q_j + \dots .$$

The Jahn-Teller theorem was proved by showing that for all symmetry groups except $D_{\infty h}$ and $C_{\infty v}$ there was at least one normal mode of vibration which belonged to a non symmetric representation Γ_i such that the direct product of Γ_i with the representation Γ_j of the degenerate electronic state contained the representation Γ_j. An alternative and equivalent statement is that the triple direct product $\Gamma_i \Gamma_j^2$ contains a component of the totally symmetric irreducible representation. Because of this result, a nuclear displacement Q_i along the normal mode belonging to Γ_i would lead to a non-zero matrix element of the perturbation to the Hamiltonian $(\partial H/\partial Q_i)\, Q_i$ between the components of the degenerate electronic state. Because of the initial degeneracy of the components the energy changes brought about by these matrix elements would be first-order (i.e. proportional to Q_i) so that at least one component of the state is lowered in energy.

The Jahn-Teller effect has been the subject of several reviews (see for example reference *10a*) and I shall not cover in this review the most familiar situations.

It may not at first be obvious that the Jahn-Teller theorem applies to transition states (*40*). The proof rests on the fact that the matrix element of the distortion gives a first-order change in energy and hence is linear in Q_i. In other words there must be a non-zero slope in some direction and this is incompatible with the definition of a transition point as a saddle point on the potential energy surface.

As simple examples of the use of symmetry to eliminate certain structures as transition states we can take the reactions

$$H + D_2 \rightarrow HD + D$$

and

$$H_2 + D_2 \rightarrow 2\,HD\,.$$

In the first of these an equilateral triangular transition state would, from simple molecular orbital arguments, have the electronic configuration (in D_{3h}) $(a_1')^2\,(e')$ and would be a degenerate state of E' symmetry. It is of course well known that the transition state for this reaction is linear. For the second reaction a square planar structure (D_{4h}) would have the molecular orbital configuration $(a_{1g})^2\,(e_u)^2$ and this gives singlet states A_{1g}, B_{1g} and B_{2g} none of which are degenerate and hence this is (by the Jahn-Teller theorem) an allowed structure for the transition state (see Fig. 4). The tetrahedral structure (T_d) however, would have the configuration $(a_1)^2\,(t_2)^2$ and this gives singlet states $A_1 + E + T_2$ all of which except A_1 would be subject to Jahn-Teller distortion.

The Jahn-Teller effect is associated with conical intersetions of a surface in two-dimensional coordinate space (Fig. 6). The normal mode that removes the degeneracy is itself doubly degenerate, or there are two non-degenerate modes that remove the degeneracy (as in the D_4 group for example). An important property of such intersections is that if we move once around the inner or outer surface of the double cone (the angle θ in the $Q_1 - Q_2$ plane increasing by 2π) then the electronic wave function

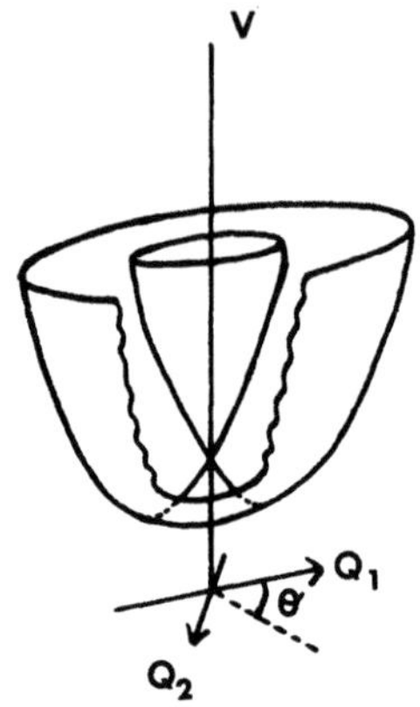

Fig. 6. The conical intersection typically associated with the Jahn-Teller effect.

changes sign (*37, 41*). This property shows that there cannot be a Born-Oppenheimer separation of the wave function into a single product of electronic and vibrational functions because if this were the case the wave function would not be single valued. The Jahn-Teller situations must therefore be represented by at least the sum of two Born-Oppenheimer terms

$$\psi_n(r_i, Q_1, Q_2)\, \psi_n(Q_1, Q_2) + \psi_m(r_i, Q_1, Q_2)\, \psi_m(Q_1, Q_2)\,. \tag{23}$$

There is a clear distinction between conically intersecting surfaces and two surfaces which touch, as in the so-called Renner-Teller effect for the degenerate states of linear molecules (*42, 43*) (Fig. 7). For such a situation the electronic wave function is single valued and although there may be strong mixing between Born-Oppenheimer terms, one such term can always be taken as a first approximation to any state.

The nuclear distortions associated with the Renner-Teller effect depend on the magnitude of the matrix elements of the quadratic terms in (22). This is in contrast with the Jahn-Teller effect that depends on the linear term. There is also an important intermediate situation which was first noted by *Öpik* and *Pryce* (*44*), and is now generally known as the pseudo Jahn-Teller effect. This arises from a mixing of close but non-degenerate electronic states through the linear term in (22). As the states are nondegenerate the mixing is formally second-order hence the contributions to the energies of the states will be quadratic in Q_i.

There have been numerous applications of the pseudo Jahn-Teller effect. Bader, for example, rationalized the signs of the quadratic interaction (off-diagonal) force constants in triatomic molecules (*45*). For a molecule ABA the two diagonal bond stretching force constants have the same positive value k_{AB} and there is one off-diagonal constant k', which may be positive or negative. The effective force constant for the symmetric stretching mode Q_s is $k_{AB} + k'$ and for the antisymmetric stretch Q_a it is $k_{AB} - k'$. Now a displacement along Q_s will lead to a pseudo Jahn-Teller stabilisa-

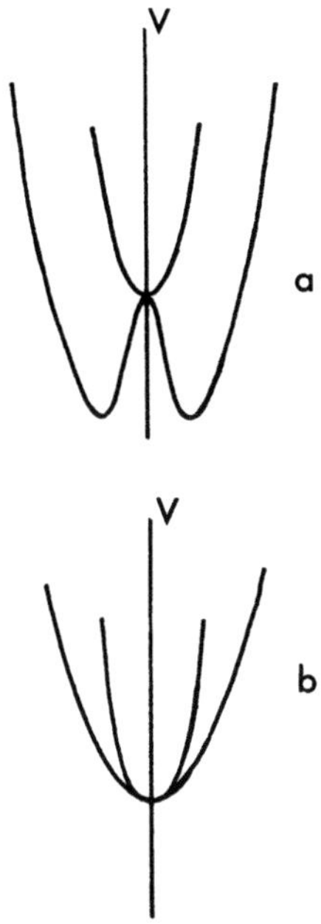

Fig. 7. Cross sections of potential surfaces for a Π state of a linear molecule showing the Renner-Teller splitting. This may (*a*) or may not (*b*) give rise to a stable bent configuration.

tion of the ground state by excited states of the same symmetry as the ground state, whereas a displacement along Q_a will lead to a stabilisation by excited states of different symmetry. The direct product of the ground and excited state representations must contain the representation of Q_a. By noting that for linear molecules such as CO_2 the first excited state to interact with the ground state was antisymmetric, Bader explained the fact that $k_{AB} - k'$ was less than $k_{AB} + k'$, that is k' was positive. For H_2O in contrast the first excited state to mix is symmetric and k' is therefore negative.

Bader (*46*) was also the first to recognize the role of the pseudo Jahn-Teller effect for predicting reaction paths and the symmetries of transition states, and in this and related papers by *Longuet-Higgins* (*47, 48*) and co-workers one can find ideas similar to those of *Woodward* and *Hoffman* (*49*) that came later but were more widely recognized. Both *Bartell* (*50*) and *Pearson* (*51*) have used the pseudo Jahn-Teller effect to explain the geometries of stable molecules and to predict reaction paths. For example, the well known *Walsh* rules (*52*) for molecular geometries were first based on simple molecular orbital arguments depending on orbital overlap and the difference between *s* and *p* atomic energies. *Pearson* (*51*) re-interpreted them from a symmetry point of view as the following example will show.

The valence molecular orbitals of D_{3h} trihydrides (AH_3) occur in order of increasing energy as

$$a_1', e', a_2'', a_2', e' .$$

For six valence electrons (e.g. BH_3, CH_3^+) the ground state is $(a_1')^2\,(e')^4$ and the lowest excited state should have $(a_1')^2\,(e')^3\,(a_2'')^1 = E''$ symmetry. The A_1' ground state would mix with E'' under an E'' vibration, but such D_{3h} structures have no normal mode of this type. This rationalizes the stable form of such species as being planar. On the other hand, for eight valence electron molecules (H_3O^+, NH_3, CH_3^- etc.) the lowest excited state has symmetry $(a_1')^2\,(e')^4\,(a_2'')^1\,(a_1')^1 = A_2''$ which mixes with the A_1' ground state by an A_2'' mode which is the out of plane bending vibration: thus such molecules should be, and are, pyramidal.

In addition to the use of symmetry to rationalize the structures and force fields of stable molecules and transition states, there has in recent years been considerable interest in its use for predicting the course of chemical reactions. These predictions are based upon the correlation of the electronic states or molecular orbitals of reactants and products through possible intermediate nuclear configurations (*49, 53–55*). Two of the standard examples will illustrate the power of the arguments. A third example will show the limitations.

We have already noted earlier that a square planar structure is allowed by the Jahn-Teller theorem as a possible transition state (Fig. 4) for the reaction

$$H_2 + D_2 \rightarrow 2\,HD\,.$$

Figure 8 shows the correlation of the lowest four molecular orbitals (approximately linear combinations of the four hydrogenic 1 *s* atomic orbitals) along a possible reaction path assuming that this has D_{2h} symmetry in intermediate configurations. It is seen that orbitals of b_{3u} symmetry which are bonding for the reactants correlate with antibonding orbitals of the products. One can therefore expect a large

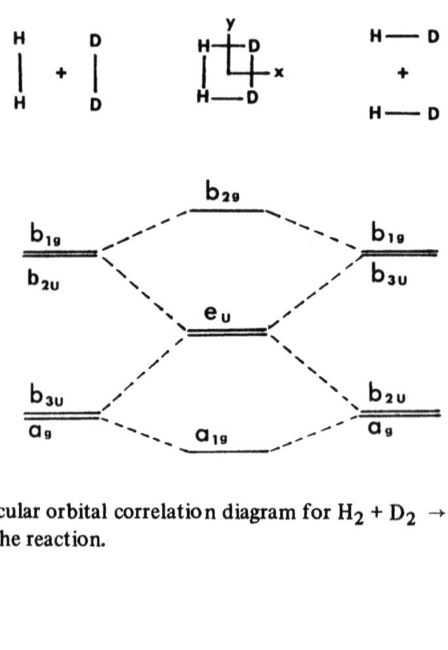

Fig. 8. Molecular orbital correlation diagram for $H_2 + D_2 \rightarrow 2\,HD$ assuming D_{2h} symmetry throughout the reaction.

increase in energy of the highest occupied orbital in the transition state and hence an increase in the ground state energy. The ground state throughout the reaction would have A_g symmetry in $D_{2h}(B_{1g}$ in $D_{4h})$ and would arise from the orbital configurations

$$\lambda(a_g)^2 (b_{3u})^2 + \mu(a_g)^2 (b_{2u})^2$$

where $\lambda = 1$, $\mu = 0$ for the reactants, $\lambda = \mu$ for the transition state and $\lambda = 0$, $\mu = 1$ for products. The increase in energy of the ground state follows in general terms from the change in orbital energies as shown in Fig. 8. Accurate calculations (*56*) give an energy for the D_{4h} structure which is 614 kJ/mole. This is greater than the dissociation energy of H_2 (109 kcal/mole) and considerably greater than the estimated bimolecular activation energy (< 200 kJ/mole) although for this reaction there is some controversy over the interpretation of kinetic results (*57, 58*).

For a second example I take the ring opening of butene to *cis* butadiene. Considerations of symmetry allow a distinction to be made between paths which are distinguished by the modes of rotation of the CH_2 groups: conrotatory and disrotatory as shown in Fig. 9. In the conrotatory mode the molecule retains a 2-fold axis of symmetry so that orbitals or states can be characterized by the symmetry labels A or B, and in the disrotatory mode there is a plane of symmetry so that the symmetry labels A' and A''.can be used.

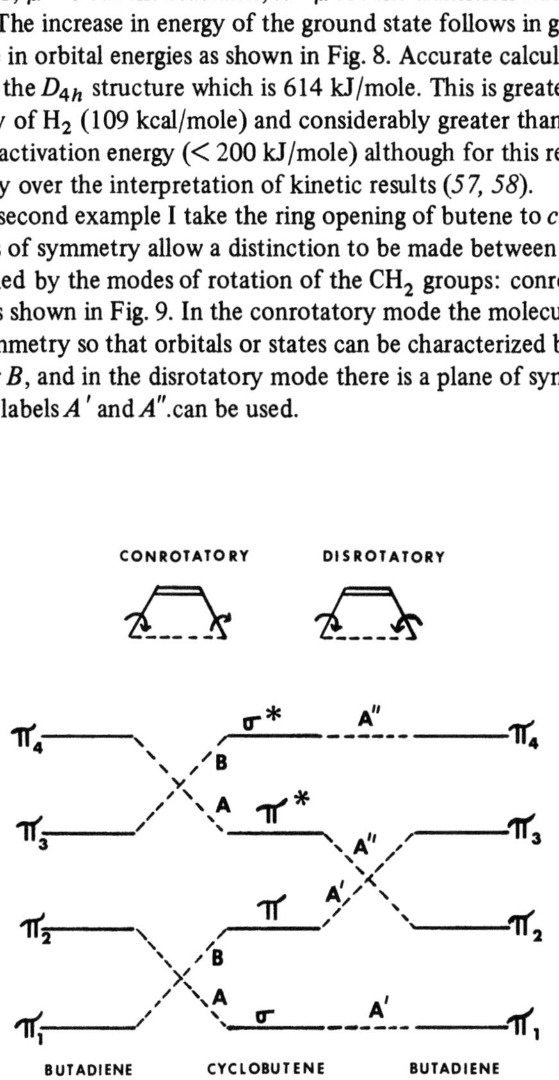

Fig. 9. MO correlation diagram for the ring opening for cyclobutene to *cis*-butadiene assuming conrotatory or disrotatory motion of the terminal CH_2 groups.

We concentrate our attention on just four molecular orbitals. In the reactants these are the bonding (π) and antibonding (π^*) orbitals of the double bond and the bonding (σ) and antibonding (σ^*) orbitals of the bond that is being broken. In the product we have the four π molecular orbitals of *cis* butadiene, which in order of increasing energy we label π_1, π_2, π_3, π_4.

Although the diagrams are schematic in the sense that there is no attempt to be quantitative in the energy scale, it is clear that the ground state of the reactant will go more easily into the ground state of the product by a conrotating mode whereas the first excited ($\pi \rightarrow \pi^*$) state will go more easily into the first excited state of the product by a disrotating mode. This is in accord with experimental findings on such structures.

Finally I take the example of the reaction

$$CH_2 + H_2 \rightarrow CH_4$$

part of whose surface has been studied theoretically using approximate ab-initio calculations (*59*). There are two low-lying electronic states of methylene, a triplet (lowest) and a singlet, in both of which the molecule is non-linear. The surfaces were first studied under the assumption that the intermediate structures had C_{2v} symmetry throughout (i.e. H_2 approaching side-on along the 2-fold axis of the CH_2) and for both the singlet and triplet surfaces the correlation diagrams showed a crossing of occupied and unoccupied orbitals analogous to that in Fig. 8. Both reactions were therefore formally symmetry forbidden and in both cases activation barriers to the reaction were found. However, by lowering the symmetry of the intermediate structures (to C_2 or C_s) the non-crossing rule was brought into play, that is, orbitals which crossed in C_{2v} did not cross in the lower symmetry. Both reactions were therefore formally symmetry allowed and the barriers to reaction were substantially removed and possibly would have been found to disappear completely if the full surface had been studied. The lesson to be learnt from this is that if there are barriers to reaction that arise from symmetry restrictions on the intermediate structures then one must expect the actual path for the reaction to go via lower symmetry configurations and a "symmetry forbidden" reaction may in fact proceed quite easily.

Consideration of symmetry may sometimes rule out a structure as being a transition state for a particular reaction even if it is allowed by the Jahn-Teller theorem. For example, transition states with D_{3h} symmetry were suggested by *Clyne* and *Thrush* (*60*) for reactions such as

$$NO + O_2 \rightarrow NO_2 + O\,.$$

Bader (*46*) considered the implications of the pseudo Jahn-Teller theorem for this and deduced that the bond stretching E' mode would be the most likely reaction coordinate, and one can see from Fig. 10 that this would lead to the required products if continued to dissociation.

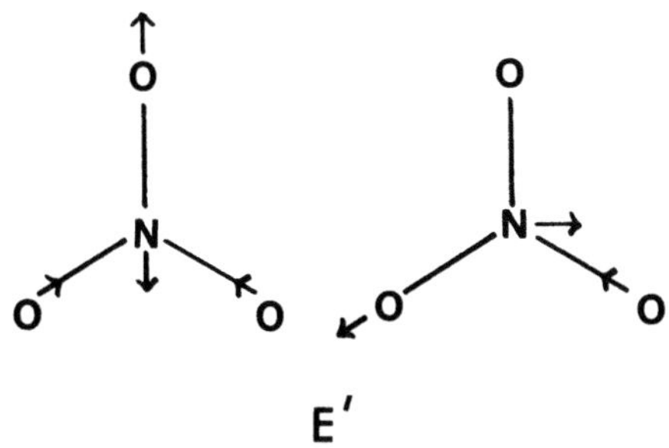

Fig. 10. One of the E' modes of vibration of a D_{3h} structure for NO_3.

However, a reaction coordinate can never be parallel to a degenerate normal mode in a transition state as this would imply that there would be more than one direction of negative curvature in the transition state. In such a situation one can always find a lower potential energy path that goes around the "hill" *(61)*. The $NO_3(D_{3h})$ structure may be a stable intermediate on the path of the reaction but it cannot be the transition state.

By considering the symmetry of the normal modes of transition states *Murrell* and *Laidler* showed that problems encountered when calculating the statistical factors of transition states (which are needed to calculate the partition function in transition state theory) were associated with configurations of too high a symmetry to be transition states *(61, 62)*.

Stanton and *McIver* *(63)* have given a more detailed discussion of the symmetries of transition states from the standpoint of the Murrell-Laidler Theorem.

We return finally in this section to the question of avoided crossings in potential energy surfaces. We have already noted in section 2 that diabatic potentials may be a more suitable basis for a dynamical calculation than adiabatic. By expanding on the relationship between the two we will be able to introduce the concept of analytic continuation of adiabatic surfaces into complex coordinate space which has proved useful for some dynamical problems.

Suppose V_a and V_b are diabatic surfaces which cross at some configuration Q_i^0 and let $\omega/2$ be the matrix element of the electronic Hamiltonian between the two diabatic states. In other words, the adiabatic curves are the eigenvalues of the potential energy matrix.

$$\begin{pmatrix} V_a & \omega/2 \\ \omega/2 & V_b \end{pmatrix}$$

which are

$$V = \tfrac{1}{2}(V_a + V_b \pm [(V_a - V_b)^2 + \omega^2]^{1/2}). \tag{24}$$

This assumes that there is no mixing with other adiabatic states in this region where $V_a \approx V_b$.

The adiabatic curves will cross only if

$$(V_a - V_b)^2 + \omega^2 = 0 \tag{25}$$

which for real quantities is only true if $V_a = V_b$ and $\omega = 0$. However, if we allow for these quantities to be complex then we have the possibility of crossing at the points where

$$(V_a - V_b) = \pm i\omega \tag{26}$$

or, if we equate real and imaginary parts, we must satisfy the conditions

$$(V_a - V_b)_R = \mp \omega_I , \quad (V_a - V_b)_I = \pm \omega_R . \tag{27}$$

For a diatomic molecule with the one internal degree of freedom (Q) the two Eqs. (27) will be satisfied at two points in the complex plane of Q. For example, suppose ω is real and $V_a - V_b$ is linear in Q,

$$V_a - V_b = A(Q - Q^0) \tag{28}$$

then the equations are satisfied at the points

$$Q = Q^0 \pm i\omega/A . \tag{29}$$

We have therefore shown that adiabatic surfaces can be said to cross off the real coordinate axes, and indeed if the classical equations of motion are solved in complex coordinate space then it is possible to simulate non-adiabatic processes. This can be considered as the basis of the *Stuckelberg* semi-classical approach to non-adiabatic transitions in atom-atom collisions (*64*) and it has been recently extended to more degrees of freedom (*65*). Moreover the actual form of potential surfaces in the complex plane has been obtained by direct calculation (*66*).

7. Calculation of Potential Energy Surfaces: General Principles

If we want to know more than the qualitative features of a potential energy surface then we must turn to a solution of the electronic Schrödinger equation or to the analysis of experimental data. In the latter area the most detailed information about a surface will be obtained from scattering experiments or spectroscopy and in general this involves a process of fitting parameters in some assumed form of the potential so that the experimental results are reproduced. In other words, even if a potential energy surface is to be deduced entirely from experimental data some theoretical knowledge of the general analytical form of the surface is usually required. For this reason I will consider the problems involved in obtaining theoretical surfaces before dealing with methods of deducing them from experimental data.

The most efficient method of obtaining approximate eigenfunctions of the electronic Hamiltonian (10) depends on the "strength" of the interaction between the atoms. This is a qualitative concept but emphasizes the fact that the best method of calculating intermolecular forces, which are generally considered as weak interactions, may be quite different to the best method of calculating valence forces which are strong interactions.

In calculating a potential energy we are not interested in the total energy but only in the energy relative to that of the separated atoms. It may be that this relative energy is best obtained by calculating a total molecular energy and subtracting the calculated separate atom energies, and this is usually the case with strong interactions. However, particularly for weak interactions there are obvious advantages in attempting to calculate the interaction energy directly.

It is beyond the scope of this review to cover in depth either valence theory or the theory of intermolecular forces and I shall only attempt to deal with some general principles of both which appear to be important for an understanding of potential energy surfaces. Before dealing separately with weak and strong interactions, there is one point they have in common and that is the increasing computational effect that is required as the number of internal coordinates increases.

It is well-known in valence theory that the computational effort required to obtain a given accuracy in the electronic energy increases rapidly as the number of electrons increases. One often refers to the N^4 explosion, N^4 being the order of the number of integrals that have to be evaluated if molecular wave function are constructed from N one-electron atomic basis functions. N must be roughly proportional to the number of electrons. However, N^4 is only a rough guide to the computational effort involved and the precise form depends on the accuracy to which the calculation is to be taken. For example, calculations have produced a potential energy curve for the ground state of H_2 which is comparable in accuracy with the best experiments over its whole range (*68*), but this cannot be said of the next smallest stable molecule LiH. On the other hand, if one is prepared to make reasonable assumptions about the passive role of inner shell electrons, then a calculation on $SiBr_4$ should take no longer than one on CF_4.

The increased computational time that has been referred to above is for a single point on the potential energy surface, and a more serious problem to be faced is the increase in the number of points necessary to obtain a coverage of the whole surface when the number of internal coordinates is increased.

Suppose we require evaluation at X points in one dimension to obtain a representation of a potential curve, then for k internal coordinates we require X^k points. The time taken to make one evaluation will in general also increase with k by virtue of the fact that the number of electrons is roughly proportional to the number of atoms (n).

The total time will be approximately proportional to

$$X^k n^4 .$$

Thus the ratio of time taken to calculate the surface for a triatomic molecule to that for a diatomic is

$$X^2 (3/2)^4 .$$

For $X = 4$, a very modest demand, this ratio is 81 so that even if we accept some saving from the carrying over of information from one point to another in the calculation, the outlook for accurate calculations on complete surfaces of polyatomic molecules is poor in the forseeable future.

The above prognosis gives weight to the idea that one should obtain as much information as possible about the form of surfaces from experimental data and to use theoretical methods only in those areas where experiment is lacking but where knowledge is vital – say in the region of transition states. I shall return to this idea in the last section.

8. Calculations in Regions of Strong Interactions

The so-called Hartree-Fock (HF) limit is important both conceptually and quantitatively in the quantum mechanical theory of many-body interactions. It is based upon the approximation in which one considers each particle as moving in an effective potential obtained by averaging over the positions of all other particles. The best energy calculated from a wavefunction having this physical significance is called the Hartree-Fock energy and the difference between this and the exact solution of the non-relativistic wave equation is called the correlation energy.

I have emphasized the HF wavefunction at the beginning of this section because it is much easier to obtain a wavefunction which approaches the HF limit than it is to obtain one that approaches the exact limit. For this reason the great majority of molecular calculations have been done within a framework which has the HF solution as a limit. Unfortunately for atomic interactions the changes in correlation energy with interatomic distance may be comparable in size to changes in the HF energy and in some regions of the surface, notably the weak interaction region, they may be considerably larger.

The most common method of going beyond the HF limit is through configuration interaction (CI). The HF wavefunction represents a definite assignment of electrons to orbitals (that is, it is describable by a single electron configuration) and to form a CI wavefunction we mix basis states having different configurations.

As an example we can take the $H_2 + D_2$ system whose correlation diagram is shown in Fig. 8. The HF wavefunction will correspond to the configuration $(a_g)^2 (b_{3u})^2$ for reactants or $(a_g)^2 (b_{2u})^2$ for products. A simple CI wavefunction will take a mixture of the two, and we have already noted that at the transition state there must, by symmetry, be an equal mixture. This example illustrates one of the most common situations where it is most important to have a wavefunction which allows for correlation to obtain a satisfactory picture of the potential energy surface. In general, if in the HF approximation there are excited configurations which are low in energy, or even degenerate with the HF solution, then the correlation energy will be large. Failure to allow for correlation in the $H_2 + D_2$ surface would lead to an overestimate of the activation energy by approximately 20% or 120 kJ/mole (taking the correlation energy as approximately half the splitting of states arising from the degenerate configurations in the transition state (*56*)).

In the regions of potential energy surfaces corresponding to stable molecular structures the correlation energy is often approximately constant. This arises because there is usually a well-defined ground state configuration with a large energy gap below this and excited configurations. Thus the geometries and force fields of stable molecules are often reproduced quite well by calculations in the HF limit and even self-consistentent-field molecular orbital (SCF MO) calculations which are far from the HF limit often give good results.

Table 1. Computed and experimental equilibrium geometry and harmonic force constants for H_2O (all results in atomic units). The results of reference 70 were obtained by fitting 36 points on the surface to a general quartic force field the first terms of which were

$$f_r(\Delta r^2 + \Delta r'^2) + 2 f_{rr'} \Delta r \cdot \Delta r' + f_\alpha \Delta\alpha^2 + 2 f_{r\alpha}(\Delta r + \Delta r') \cdot \Delta\alpha .$$

The results of reference 69 were obtained by fitting only these quadratic terms to a nine point grid.

	r_{OH}	α	f_r	$f_{rr'}$	f_α	$f_{r\alpha}$	Ref.
Minimal basis SCF MO	1.871	100.3	0.256	– 0.09	0.124	0.020	(*69*)
Near HF	1.760	106.08	0.3145	– 0.0021	0.089	0.015	(*70*)
CI	1.8004	104.93	0.2850	– 0.0027	0.085	0.016	(*70*)
expt	1.8089	104.52	0.2709	– 0.0038	0.085	0.015	(*71, 72*)

Table 1 illustrates this for the case of H_2O. The CI calculation gives results in good agreement with experiment, and this holds also for the cubic and quartic anharmonic terms in the force field. The near HF results are less good but may still be regarded as acceptable if one realises that for many molecules the complete harmonic force field is still not known with accuracy. The over-estimate of the diagonal force constant f_r is generally associated with the fact that the HF wavefunction has an incorrect dissociation limit so that the potential well is too deep with respect to its calculated atomic fragments.

The minimal basis SCF MO calculations show appreciable errors in the force field and equilibrium geometry, although if one recognizes that very small energies are associated with angle-bending, the fact that the equilibrium bond angle is in error by only 5° is perhaps gratifying. It is important to note the limitations of minimal basis calculations because these represent by far the largest volume of ab-initio calculations in the literature and nearly all semiempirical calculations are based upon this approximation.

It could reasonably be argued that for the purpose of obtaining information about potential energy surfaces we will not often need to do calculations in the region of stable molecular configurations because spectroscopic data will usually provide more accurate information. It is in regions far from equilibrium that calculations will prove of most value, and for these we have already noted that the HF approximation may be poor.

At the present time there have been few ab-initio calculations on polyatomic potential surfaces for which a claim of quantitative accuracy can be made. The classic example of the H_3 surface stands out in this field as a longstanding challenge to theoreticians, and the best calculations on this are thought to be within 12 kJ/mole of the exact surface overall, and better than 4 kJ/mole for linear configurations which encompass the transition state (*73–75*).

Other small triatomics for which quite accurate calculations have been made with a good coverage of the complete surface are H_3^+ (*76–78*), He + H_2 (*79, 80*),

$Li + H_2$ (*81, 82*), $F + H_2$ (*83*) and $H + F_2$ (*84*). References to other polyatomic surfaces studied by ab-initio methods, including correlation, have been given by *Balint-Kurti* (*85*) and by *Bader* and *Gangi* (*85a*).

There are several ways of attacking the correlation problem within the CI framework. These lie between the extremes of having a very large number of configurations with no attempt to choose them in an optimal sense – current computer technology can handle several thousand configurations – or a small number of configurations carefully chosen. The latter procedure will, of course, lead to wavefunctions which are more amenable to physical interpretation, and although that facility might be said to be irrelevant to the present problem, one should not completely dismiss physical models as being guides to good wavefunctions. The extreme of short but accurate CI expansions occur within the framework of the multi-configuration (MC) SCF calculations. In a normal CI method the orbital basis is first determined by a SCF calculation and these orbitals are then used to construct the excited configurations. The convergence of the CI expansion may be improved by first making equivalent transformations of these orbitals, so that they approach the so-called natural orbitals (*86–88*), but even with this the method commonly needs several hundred configurations to obtain accurate energies. In the MC SCF method, however, the orbitals are chosen not to be optimum for the SCF wavefunction but for the whole CI expansion. To reach this objective is in practice a time-consuming procedure, but it may be carried through for a small number of configurations. It is at this point that a physical model is important to determine which configurations to use.

An excellent example of the MC SCF procedure is given in a paper on F_2 by *Das* and *Wahl* (*89*). This amongst diatomic molecules is one of the standard theoretical challenges because in the HF limit the energy of F_2 is greater than twice the HF energy of the fluorine atom – in other words, there is no binding energy.

The MC SCF wavefunction is built up in the following stages:

(i) The SCF procedure is carried out to near HF accuracy.

(ii) Configurations are added which will make the wavefunction extrapolate to the correct dissociation limit and the orbitals are re-optimized.

An incorrect dissociation limit is a common failure of SCF MO wavefunctions (as we already noted for H_2O). Thus for H_2 the SCF MO wavefunction (σ_g^2) leads to a dissociation limit which is an equal mixture of atoms and ions because there is no correlation between the two electrons (there is an equal chance of finding the two electrons on the same atom and on different atoms). The addition of a configuration (σ_u^2), where σ_u is the lowest energy unfilled molecular orbital, removes this error, and in the dissociation limit the wavefunction has to be an equal mixture of σ_g^2 and σ_u^2. Thus a wavefunction that stops at this limit is called an optimum double configuration (ODC) function.

(iii) Configurations are added which introduce inter-atomic correlation. There are several categories of these which have in common the property that their contribution goes to zero on dissociation. The orbitals will again be optimized.

(iv) Configurations are added which introduce intra-atomic correlation and ideally the orbitals should again be optimized. The contribution from these will not go to zero at the dissociation limit but will change, hopefully slowly, with the internuclear separation.

The four steps outlined above appear to be a good general approach to diatomic potentials, as can be seen from the F_2 results shown in Fig. 11. The first two steps are essential to obtain a physically acceptable surface; the third step may add appreciably to the binding energy particularly for weak bonds, and the last step, which is the most tedious, is hopefully not essential for many molecules. It has however still to be established that good MC SCF wavefunctions for polyatomic molecules may be constructed by similar arguments.

It is clear that with the present stage of theory and computer technology accurate energies by ab-initio calculations can only be expected in the near future for small

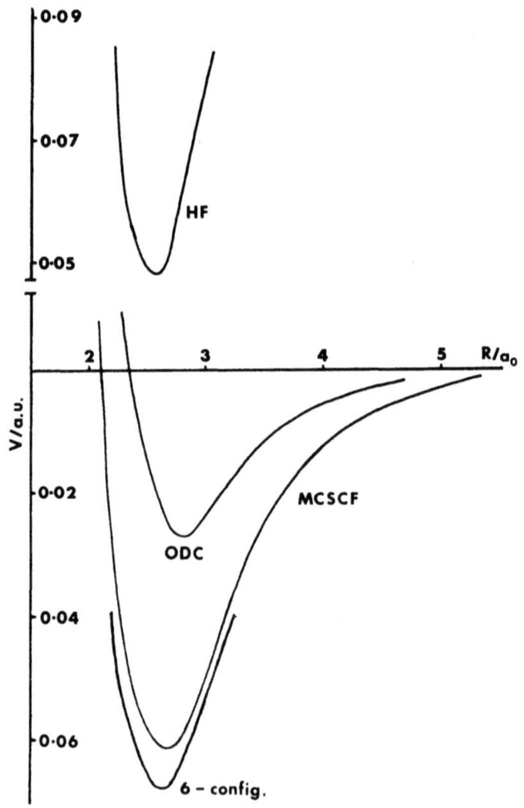

Fig. 11. Calculations on F_2 (*89*): Hartree-Fock (HF), optimized double configuration (ODC) and multiconfiguration self-consistent-field (MCSCF). The latter is close to the experimental curve. A 6-configuration curve is also shown which excludes intra-atomic correlation terms.

numbers of light atoms and with a modest coverage of the whole surface. There is therefore scope for quicker qualitative calculations which will at least give some guide to the characteristic features of a surface. The history of semiempirical methods goes back to the early years of quantum mechanics. *London* (*4*) himself proposed an empirical formula for a triatomic potential energy based upon the Coulomb (Q) and exchange (J) integrals that appeared in the original Heitler-London treatment of the hydrogen molecule.

$$E = Q_{ab} + Q_{bc} + Q_{ca} - \sqrt{\tfrac{1}{2}}\,[(J_{ab} - J_{bc})^2 + (J_{bc} - J_{ca})^2 + (J_{ca} - J_{ab})^2]^{1/2} \tag{30}$$

This expression has seen many developments through the years and has evolved into the so-called London-Eyring-Polanyi-Sato (LEPS) surface in which expression (30) is multiplied by an empirical factor $(1 + k)^{-1}$ which is supposed to take account of overlap effects (*90*). The coulomb and exchange integrals are calculated from the singlet and triplet potential curves of the diatomics, given by the expressions

$$\begin{aligned} {}^1E_{ab} &= (Q_{ab} + J_{ab})\,(1 + k)^{-1} \\ {}^3E_{ab} &= (Q_{ab} - J_{ab})\,(1 - k)^{-1} \end{aligned} \tag{31}$$

with the singlet curves taken to be Morse functions

$${}^1E_{ab} = D_e[e^{-2\beta(r-r_e)} - 2\,e^{-\beta(r-r_e)}] \tag{32}$$

and the triplets a repulsive analogue

$${}^3E_{ab} = D_e[e^{-2\beta(r-r_e)} + 2\,e^{-\beta(r-r_e)}]\ . \tag{33}$$

The parameter k is then chosen to give the LEPS surface some desired property, for example a specific activation energy (*91–93*). Further empirical parameters can give the surface added flexibility (*94*).

It might be argued that the LEPS method is purely empirical in that none of the quantities that enter it are calculated and only the general form has some theoretical foundation. It has nevertheless been widely used in dynamical calculations to investigate the influence of certain features of the surface, such as the position and height of the transition state.

The London equation has in addition been the progenitor of semi-empirical (or semi-theoretical) valence-bond methods of which Moffitt's method of atoms in molecules (*95*) and Ellison's method of diatomics in molecules (*96*), the latter not in fact being a direct generalisation of the former, are the most important. It is beyond the scope of this article to give the details of these methods and their modifications and the reader is referred to other reviews that encompass them (*85, 97*). It is however important to emphasize that they work by using relatively simple polyatomic wavefunctions and introducing corrections to the resulting Hamiltonian matrix elements by first identifying the errors which such functions give for the atomic or diatomic dissociation fragments and correcting for them. Thus they use atomic or diatomic data which may be empirical or from accurate calculations.

We should also not leave this section without referring to the widely used Porter-Karplus surface for H_3 which was constructed by semi-empirical valencebond methods and which agrees quite well with the best ab-initio surfaces (*98*).

Within the framework of molecular orbital theory there have been developments of empirical theories ranging from Huckel π-electron theory to the more sophisticated all-valence-electron zero-overlap theories. A review of the main developments in the field is given in reference (*99*). Most of the semi-empirical theories have been parameterized from the properties of atoms and stable molecules. For some, such as the CNDO model of *Pople, Santry* and *Segal* (*100*), and its modifications, the parameterization has the aim of reproducing the results of ab-initio SCF MO calculations. They must therefore face the same limitations regarding the general applicability to potential energy surfaces as the ab-initio SCF MO method. In others, notably the MINDO model of Dewar and co-workers (*101*), the parameterization has been based upon experimental quantities, and insofar as some of these are directly connected with features of the potential energy surface such as equilibrium configurations, force-constants and dissociation energies, they might be expected to go beyond the SCF MO limits. In other words, they have been parameterized to give correct potential energies for some selected standard molecules.

The complexity of the parameter-fitting procedure in the MINDO models can only be appreciated by a detailed study of the inherent assumptions. It is perhaps indicative to say the only molecular integral that is calculated exactly from the basis atomic orbitals is the overlap integral, all others being approximated or given empirical values. The repulsion potential of the atomic cores is, for example, one of the critical functions in the theory. There has to the present time been four distinct versions of the MINDO parameterization, and the latest (MINDO/3) (*102*) is said to remove certain deficiencies in the earlier versions (such as the prediction that H_2O was linear and the underestimation of the strain energies in small ring hydrocarbons).

The MINDO programme has been developed mainly to handle small and medium sized organic molecules, and it is in this field that a large number of potentialenergy studies have been made. Typical of these is the following study of the conversion of Dewar benzene to benzene (*103*).

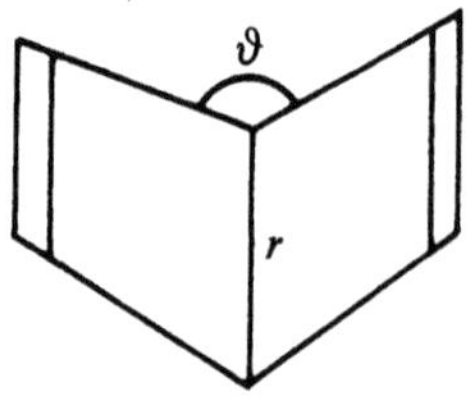

The potential surface was calculated as a function of the inter-annular bond length (r) and the angle between the planes of the four membered rings (ϑ) with all other coordinates being optimized. With the assumption of D_2 symmetry (three mutually perpendicular 2-fold axes) or higher, the reaction was found to be "symmetry forbidden", that is there was a crossing of bonding and antibonding orbitals on going from reactants to products. The corresponding activation energy was found to be 174 kJ/mole (observed 96 kJ/mole). Lowering the symmetry led to a transition state in which the inter-annular bond had been stretched to 2.04 Å, but was not symmetrically placed with respect to the centre of the plane defined by the four "olefine" carbon atoms. The activation energy was reduced to 150 kJ/mole. This result is analogous to that already quoted for the system $^1CH_2 + H_2$ in which a barrier that is associated with orbital crossing is circumvented by taking a path through lower symmetry configurations.

Although one must treat the results of calculations such as that described above with caution I believe they are capable of giving a qualitative picture of the reaction paths of rather complicated organic systems. The question at issue is likely to be how much more information do they give than even more qualitative theories which perhaps bring out only the symmetry aspect, and how reliable are quantities such as activation energies likely to be: the test will be by results rather than by theoretical justification of the models.

9. Calculations in the Weak-Interaction Region

The weak interaction region can be defined as one for which the total electron density is approximately equal to the sum of the densities of the separate interacting particles. Whether one uses a direct variational method to calculate the energy or a perturbation expansion it is found that good results are only obtained if the wavefunctions for the interacting particles give accurate values for these atomic densities.

For the strong interactions dealt with in the last section, interatomic potentials are determined by overlap-dependent terms which have an inverse exponential variation with internuclear distance (R). At large separations the interactions are independent of overlap and may be represented by inverse powers of R. There will be between these two an intermediate region, which encompasses the van der Waals' minimum for non-bonded interactions, and we will take this to be contained in our definition of a weak-interaction.

There is no hard boundary between the overlap dependent and independent regions. One can only ask at which point the overlap dependent energies become comparable to the overlap independent. If we take H_2 as an example, the separation of the lowest singlet and triplet state depends primarily on overlap (in any model based upon atomic wavefunctions) whereas the absolute energies contain overlap independent terms. Table 2 shows these as a function of interatomic distance (*104, 105*) with the Coulomb (overlap independent) and exchange (overlap dependent) terms being given by

$$\begin{aligned} E_{\text{Coul.}} &= \tfrac{1}{2}({}^1E + {}^3E) - 2E_H \\ E_{\text{ex}} &= \tfrac{1}{2}({}^3E - {}^1E) . \end{aligned}$$

The van der Waals' distance for the triplet state is close to 8 a_0, and it is seen that at shorter distances than this the overlap terms become dominant.

A theoretical calculation ot the long-range energies[4]) is usually based on perturbation theory. If electron exchange can be ignored, then the product of eigenfunctions of separate systems A and B form a basis of zeroth order states according to the following equations

$$\begin{aligned} H_A(i)\,\psi_{ar}(i) &= E_{ar}\,\psi_{ar}(i) \\ H_B(j)\,\psi_{bs}(j) &= E_{bs}\,\psi_{bs}(j) \\ H &= H_A(i) + H_B(j) + U(i,j) \\ \psi_{rs} &= \psi_{ar}(i)\,\psi_{bs}(j) . \end{aligned} \tag{34}$$

[4]) We ignore a discussion of the so-called "very long-range energies" which are obtained within a field theory framework (see reference 106). They have little significance for chemists.

Table 2. Comparison of Coulomb and exchange energies (cm^{-1}) for H_2 (*104, 105*).

R/a_0	$-E_{coul}$	E_{ex}
4	1070	2523
6	70	110
8	8.0	3.6

The long-range energy arises from the first-order term

$$E_{rs}^{(1)} = \langle \psi_{rs} | \mathrm{U} | \psi_{rs} \rangle \tag{35}$$

and the second-order energy which may be expressed as

$$E^{(2)} = \sum_{t,u} \langle \psi_{rs} | \mathrm{U} | \psi_{tu} \rangle^2 \, (E_{rs}^{(0)} - E_{tu}^{(0)})^{-1} \tag{36}$$

plus higher orders which for our purpose are of no importance.

These expressions are put in a more convenient form by defining one-electron densities as follows

$$\rho_{rt}^{a}(1) = N_i \int \psi_{ar}(i) \, \psi_{at}(i) \, d\tau_{i \neq 1} \tag{37}$$

where N_i is the number of electrons on system A. The integration is over all coordinates except the space of electron 1. When $r = t$ this density is the one-electron density in the state ψ_{ar}. If $r \neq t$ the expression defines the transition density between the two states (*107*). The matrix elements in both 35 and 36 can be written as integrals involving these densities. For example, if $r \neq t$ and $s \neq u$ the only part of U which contributes to the matrix elements of (36) is the electron repulsion $\sum_{i,j} r_{ij}^{-1}$, and this gives a term, known as the dispersion energy, having the form

$$E_{\mathrm{disp}} = \sum_{t}{}' \sum_{u}{}' \left[\int \rho_{rt}^{a}(1) \, r_{12}^{-1} \, \rho_{su}^{b}(2) \, dv_1 \, dv_2\right]^2 (E_{rs}^{(0)} - E_{tu}^{(0)})^{-1}. \tag{38}$$

The dispersion energies can be represented by a multipole expansion beginning in R^{-6}. This follows because the leading moment of a transition density is a dipole and a dipole-dipole interaction when squared varies as R^{-6}. An expansion terminated at R^{-10}, which for atoms has only the even terms

$$C_6 R^{-6} + C_8 R^{-8} + C_{10} R^{-10}\,, \tag{39}$$

is generally a good representation of the dispersion energy up to the van der Waals' distance, but at shorter distances the expansion diverges and the full expression (38) must be used (*108–110*). The coefficients $C_6 - C_{10}$ are known quite accurately for several systems such as the rare gas dimers (*111*).

Those terms in the summation (36) for which $t = r, u \neq s$ or $t \neq r, u = s$, represent the energy arising from one atom or molecule being polarized by the other. This is called the induction energy. For neutral systems it is usually smaller than the dispersion energy. For details of the methods of calculating long-range interactions the reader is referred to Ref. (*108–112*).

The development of a perturbation expansion in the region where electron exchange cannot be neglected has been an area of considerable theoretical interest over the past decade. Several methods have been proposed but none are without difficulty.

The problem is to have a perturbation solution to the Schrödinger equation with the restriction that the wavefunction shall be antisymmetric to exchange of the electrons between the interacting systems. The simple product functions ψ_{rs} (in 34) do not satisfy this restriction individually and if we define a set of antisymmetrized products

$$\Psi_{rs} = \mathrm{A}\,\psi_{rs} \tag{40}$$

then this is found to be *overcomplete*. In other words, any function Ψ_{rs} can be expanded in terms of the other members of the set. This is simply proved by noting that the set ψ_{rs} is a complete orthogonal set hence we can make the expansion

$$\Psi_{rs} = \sum_{t,u} \langle \Psi_{rs} | \psi_{tu} \rangle \, \psi_{tu} \tag{41}$$

By applying the antisymmetrizer A to both sides and noting that the effect of A on Ψ_{rs} is just to multiply it by a constant (λ say) we obtain the required proof

$$\Psi_{rs} = \sum_{t,u}{}' \langle \Psi_{rs} | \psi_{tu} \rangle \, \Psi_{tu} / (\lambda - \langle \Psi_{rs} | \psi_{rs} \rangle) \,. \tag{42}$$

If the set (*40*) is truncated, then it can be used as a basis for a variational wavefunction, although its rate of convergence for this task may not be high. An alternative procedure is to use the zeroth-order antisymmetrized function for the state in question, and expand the changes in this wavefunction in terms of single product functions (*113*). This leads to perturbation expressions for the energy which compare favourably with the best variational wavefunctions but the resulting wavefunction is not exactly antisymmetric. For a review of perturbation expansions of the intermediate range of interaction see Ref. (*105, 114* and *115*).

The results of calculations over the past few years for the ground states of rare-gas dimers whether by variational or perturbation methods, show that the weak interaction potential is quite well reproduced by combining a first-order energy which is the expectation value of the antisymmetrized product of the two atom ground state wavefunctions, plus a dispersion energy which is added as the multipole expansion. This may be in part due to a fortuitous cancellation of an overestimate of the dispersion energy with an underestimate of higher order exchange terms. To generalize this result to the interaction of charged or polar species one would at least have to add an inductive energy as further terms in the R^{-n} expansion. Only a few of the references on these variational (*116, 117*) or perturbation (*118*) calculations are given here. We also note in passing some quite successful calculations based upon the assumption that the total density is equal to the atomic density and that this can be treated locally as if it were a freeelectron gas (*119, 120*).

10. Analytical Functions for Potential Energy Surfaces

For the purpose of any dynamical calculation it will generally be necessary to have the potential energy as an analytical function of the internal coordinates. This will certainly be true if the equations of nuclear motion are to be solved analytically, and even if they are solved numerically one needs a method for rapid evaluation of the potential at any point on the surface and this is only possible if an explicit analytical function is available.

There has been very little work on the form of potentials for polyatomic molecules which cover the whole space of the internal coordinates. There are however three limited regions which have been rather fully investigated and I shall deal with these first.

i. The diatomic molecule

The potential energy curves for the states of most diatomics are generally known with high accuracy from the analysis of spectroscopic or scattering data. Many analytical functions have been proposed which reproduce the main features of attractive and repulsive potentials and we have already mentioned the Morse potentials of equations (32) and (33) as typical of these. Comprehensive reviews of other functions have been given by *Varshni* (*121*) and *Goodisman* (*122*).

The attractive Morse potential has three variable parameters which can be fitted to the dissociation energy, the harmonic force constant and the equilibrium distance. The higher force constants are then in error, but as *Hulburt* and *Hirschfelder* (*123*) showed these can be corrected by multiplying the repulsive part of the potential by a polynomial as follows

$$V = D_e [(1 + a r^3 + b r^4) e^{-2\beta r} - 2 e^{-\beta r}] \tag{43}$$

r being the displacement from equilibrium. However, this function differs very little from the simple Morse function in the attractive branch ($r > 0$) and often shows considerable deviation from the exact potential. It has been found (*124*) that if one attempts to modify the attractive branch of the Morse potential by a polynomial in the same way then the resulting curves shows spurious turning points.

Another simple analytical function that is of comparable accuracy to the Morse function but which has been little used by spectroscopists is that of *Rydberg* (*125*).

$$V = -D_e [1 + ar] e^{-ar} . \tag{44}$$

This can be considered as the simplest member of a family of functions in which one takes a polynomial, or ratio of two polynomials, and multiplies this by an exponential to ensure the correct asymptotic limit. It has been found (*124*) that the function

$$V = -D_e\,[1 + ar + br^2 + cr^3]\,e^{-ar} \qquad (45)$$

in which the coefficients are fitted to the quadratic, cubic and quartic force constants, is the most satisfactory member of this family. It is in nearly all cases superior to the Hulbert-Hirschfelder function and shows no spurious turning points. I mention this function because of its relationship to more general polyatomic potentials which I wish to consider later.

The analysis of spectroscopic data for bound states of diatomic molecules gives accurate potential curves if one follows the semi-classical Rydberg-Klein-Rees method. For a review of this see Ref. (*126*). It is sufficient to note that this gives the two values of r as a function of potential energy by considering the dependence of the total spectroscopic energy on the vibrational and rotational quantum numbers n and J. A somewhat simpler procedure, and the only one applicable to polyatomic molecule, is to use the Dunham expansion of the potential (*127*).

$$V = \xi^2(a_0 + a_1\,\xi + a_2\,\xi^2 + \ldots) \qquad (46)$$

where

$$\xi = (r - r_e)/r_e \qquad (47)$$

from which, again using semi-classical theory, explicit relationship between the coefficients a_n and spectroscopically determined parameters may be derived. A short Dunham expansion, say up to cubic or quartic terms, diverges rapidly from the true potential, and is usually poor at displacements $\xi > \pm\, 0.5$. A modified Dunham expansion (*128*) with ξ being given by

$$\xi = (r - r_e)/r \qquad (48)$$

is much superior for positive ξ, and has the advantage of giving the correct asymptotic limit. *Bickes* and *Bernstein* (*129*) have recently shown that this potential reproduces the empirical multiparameter potentials derived from elastic scattering experiments with good accuracy. It is therefore likely to be used widely in the future.

ii. Polyatomic Potentials in the Region of Stable Molecular Configurations

The analysis of molecular vibration and rotational energy levels in terms of assumed force fields has a long history but it is only in recent years that theory and experiment have advanced to the stage where reliable anharmonic potentials have been obtained for small polyatomics.

Within the harmonic approximation the choice of a system of internal coordinates is irrelevant provided they are independent and that a complete potential function is considered[5]). For example, the vibrations of H_2O can be analysed in terms of valence coordinates (r_1, r_2, α) or interatomic coordinates (r_1, r_2, r_3) and any difference in the accuracy to which observed energy levels are fitted (considering all the isotopic species H_2O, HDO and D_2O) will be due to the neglect of anharmonic terms. If one makes the approximation of a diagonal force field so that one is comparing the two potentials

$$V = f_{11} r_1^2 + f_{22} r_2^2 + f_{\alpha\alpha} \alpha^2 \qquad \text{(valence forces)}$$
$$V = f_{11} r_1^2 + f_{22} r_2^2 + f_{33} r_3^2 \qquad \text{(central forces)}$$

then a choice can be made as to which is the better.

To obtain the anharmonic terms in the potential, on the other hand, the choice of coordinates is important (*130, 131*). The reason is that the anharmonic terms can only be obtained from a perturbation expansion on the harmonic results, and the convergence of this expansion differs considerably from one set of coordinates to another. In addition it is usually necessary to assume that some of the anharmonic interaction terms are zero and this is true only for certain classes of internal coordinates. For example, one can define an angle bend in H_2O either by a rectilinear displacement of the hydrogen atoms or by a curvilinear displacement. At the harmonic level there is no difference between the two, but one can see that a rectilinear displacement introduces some stretching of the OH bonds whereas the curvilinear displacement does not. The curvilinear coordinate follows more closely the bottom of the potential well (Fig. 12) than the linear displacement and this manifests itself in rather small cubic stretch-bend interaction constants whereas these constants are larger for rectilinear coordinates. A final and important point about the choice of curvilinear coordinates is that they are geometrically defined (i.e. independent of nuclear masses) so that the resulting force constants do not depend on isotopic species. At the anharmonic level this is not true for rectilinear coordinates as it has been shown that the imposition of the Eckart conditions, that the internal coordinates shall introduce no overall translation or rotation of the body, forces them to have a small isotopic dependence (*132*).

Force fields up to quartic anharmonic terms are now known with reasonably high accuracy for several triatomic molecules and the results shown in Table 3 for H_2O are typical. However, even for these there has had to be an assumption that some of the quartic interaction terms are zero in order that the equations from which the constants are derived shall have unique solutions. It can be seen moreover that some of the cubic and quartic terms have uncertainties which are larger than the values of the constants themselves.

[5]) The choice of three bond lengths for a linear triatomic gives some difficulty at the harmonic level because of their lack of independence.

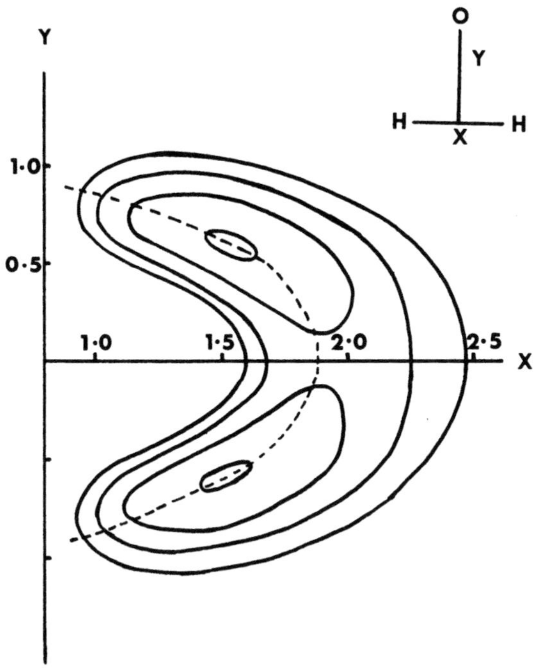

Fig. 12. Potential energy surface for C_{2v} structures of H_2O. Rectilinear displacements of the hydrogen atoms would occur along straight lines on this surface. The dotted line shows a curvilinear motion keeping the OH bond length constant. The contours (– 10, – 9, – 8, – 7 eV) have been drawn from the analytical potential of *Murrell* and *Sorbie* (*135*).

Table 3. The force field of H_2O (*131*). This is defined in the expansion of *Pliva*, but other definitions are in common usage.

$$V = \sum_{i \leqslant j} f_{ij} R_i R_j + \sum_{i \leqslant j \leqslant k} f_{ijk} R_i R_j R_k + \dots \quad (49)$$

The bracket zeros are assumed values and the standard errors are given in parenthesis.

f_{rr}	4.227	f_{rrr}	– 9.89	(.50)	f_{rrrr}	16.0	(2.6)
$f_{rr'}$	– 0.101	$f_{rrr'}$	0.13	(.75)	$f_{rrrr'}$	– 0.8	(2.9)
$f_{r\alpha}$	0.219	$f_{rr\alpha}$	0.20	(.10)	$f_{rrr'r'}$	0.1	(1.8)
$f_{\alpha\alpha}$	0.3485	$f_{rr'\alpha}$	– 0.40	(.28)	$f_{rrr\alpha}$	(0)	
		$f_{r\alpha\alpha}$	– 0.11	(.05)	$f_{rrr'\alpha}$	(0)	
		$f_{\alpha\alpha\alpha}$	– 0.146	(.017)	$f_{rr\alpha\alpha}$	– 0.35	(.02)
					$f_{rr'\alpha\alpha}$	0.30	(.25)
					$f_{r\alpha\alpha\alpha}$	(0)	
					$f_{\alpha\alpha\alpha\alpha}$	– 0.003	(.008)

Units f_{rr} md/Å, $f_{\alpha\alpha}$ md Å etc.

iii. Atom-diatom Scattering Potentials

We have already mentioned (expressions 30–33) the widely used LEPS surface for atom-diatom reactions. This may be regarded as purely empirical or semi-empirical in any modification in which some integrals are evaluated. Another system for which fairly elaborate potential functions have been used is for non-reactive atom-diatom scattering. The experiment for which the potential is designed is the change of rotational or vibrational state of a diatomic molecule by collision with a third atom, and also the quasi bound states, which may be observed spectroscopically, of "van der Waals' molecules" such as Ar–H_2 (*133*).

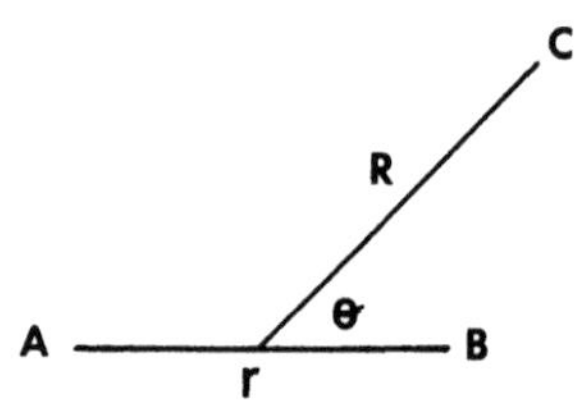

Fig. 13. Coordinates for non-reactive atom-diatom collisions. The position of *C* is measured relative to the centre of mass of *A* and *B*.

The internal coordinates usually chosen to study non-reactive atom-diatom collisions are shown in Fig. 13. If the potential is written

$$V = \sum_n F_n(r, R)\, P_n(\cos\theta) \tag{50}$$

where $P_n(\cos\theta)$ is the Legendre polynomial of degree n, then this simplifies, as far as possible, the matrix of V between rotational states of the diatomic molecule. Functions such as (50) have been fitted to points on calculated ab-initio surfaces for some systems (e.g. Ref. *79* and *82*).

The functions $F_n(r, R)$ may be taken to have the form of any diatomic potential for R, with the inclusion of an r dependence in the constants. In practice one is only interested in small displacements of r from equilibrium so that a linear dependence on r is usually sufficient to describe the important features of the surface. As an example of such a function we can take the He + H_2 potential obtained by *Gordon* and *Secrest* (*79*) of the form

$$\exp(-\alpha_0 R + \alpha_1 Rr)\,[1 + \gamma_0 r + \beta(1 + \gamma_2 r) P_2(\cos\theta)] \tag{51}$$

where r is measured from the equilibrium position of the diatomic. By symmetry there can be no P_1 term. This is by no means a potential that expresses the full flexibility of expression (50) as it uses the same exponential function for the P_0 and P_2

terms. Even so it contains five parameters that would require confirmation by analysis of experimental data.

The most extensive potential obtained so far with experimental confirmation is that of Le Roy and Van Kranendonk for the H_2 – rare gas complexes (*134*). These systems have been found to be very amenable to an adiabatic model in which there is an effective $X-H_2$ potential for each vibrational-rotational state of H_2 (c.f. the Born Oppenheimer approximation of a vibrational potential for each electronic state). The situation for $Ar-H_2$ is shown in Fig. 14, and it appears that although the levels with $H_2(\nu = 1)$ are in the dissociation continuum they nevertheless are quasi bound and give spectroscopically sharp lines.

The important point used by Le Roy and Van Kranendonk is that if the overall angular momentum state $J = 0$ is considered then the position of the levels is determined only by the P_0 terms in (50) and if $J = 2$ is considered then P_0 and P_2 are relevant. They found that a potential terminated at this point satisfactorily fitted higher J levels. For the F_n functions they chose a modified Lennard-Jones potential

$$\frac{C_m}{R_m}(1 + S_1 r) - \frac{C_6}{R^6}(1 + S_2 r) \tag{52}$$

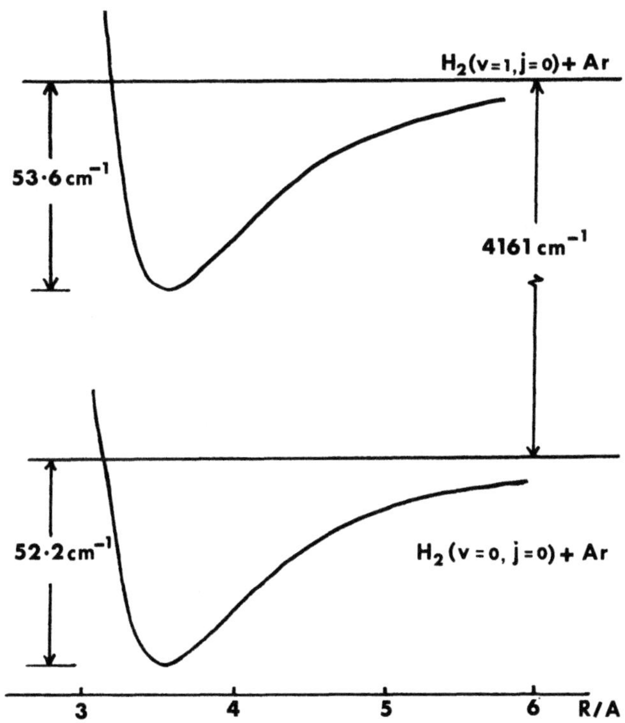

Fig. 14. Effective potentials for $Ar-H_2$ for two rotational-vibrational states of H_2.

in which C_m and C_6 are determined by the position and depth of the minimum, where m has the same value (~ 12) for both F_0 and F_2. The resulting potentials therefore have nine spectroscopically determined parameters. This is a larger number than one could expect to determine from a scattering experiment.

IV. A Spectroscopic Approach to Polyatomic Potentials with Correct Asymptotic limits

Having discussed the form of analytical potentials for three special categories of surfaces I will now describe a more general method for obtaining analytical potential energy functions for polyatomic molecules from spectroscopic data (*135*). The method is probably simpler to follow if a specific example is taken.

As an objective we wish to find a potential for the lowest singlet state of H_2O which will be in agreement with the spectroscopic properties of the molecule and its dissociation fragments and will also enable one to study the results of collisions such as

$$H_2 + O(^1D)$$

or

$$H + OH(^2\Pi)$$

In order that the potential shall be equally suitable for all configurations on the surface we choose as internal coordinates the three bond lengths. We then write the general ABC triatomic potential as

$$V(r_1, r_2, r_3) = V_{AB}(r_1) + B_{BC}(r_2) + V_{CA}(r_3) + V_I(r_1, \dot{r}_2, r_3)\,. \tag{53}$$

This involves no approximation, but we shall see later that the next and obvious step does. This is to ensure that V_I becomes zero in the dissociation limits so that the first three terms are identified with diatomic potentials.

$$\operatorname*{Limit}_{r_i \to \infty} V_I = 0\,. \tag{54}$$

Of course the triangular restriction $r_i + r_j \geqslant r_k$ implies that at the atom-diatom dissociation limit two of the three bond lengths tend to infinity.

To identify the diatomic potentials we consider the correlation rules for molecular wavefunctions. For any point on the surface there is an electronic wavefunction Ψ_{ABC} and by considering the removal of the three atoms in turn we can identify three possible dissociation limits with wavefunctions $\psi_{AB}\ \psi_C$, $\psi_{BC}\ \psi_A$, $\psi_{CA}\ \psi_B$. If we impose spin conservation on the wavefunctions then these limits for the ground state of H_2O are as follows

$$H(^2S) + OH(^2\Pi)$$

$$H_2(^1\Sigma_g^+) + O(^1D) \qquad (i)$$

or

$$H_2(^3\Sigma_u^+) + O(^3P) \qquad (ii)$$

with the appropriate spin pairing in each case.

We see that there are two possible limits for H_2 + O. We take the one of lower energy as appropriate and this will depend on the H–H separation. The $^1D-^3P$ separation of the oxygen atom is 1.958 eV and calculations by *Kolos* and *Wolniewicz* (*136*) show that the singlet-triplet separation of H_2 is equal to this at 1.669 Å. Thus the two-body potential V_{HH} that we use in (53) will be the singlet state potential for $r_{HH} < 1.669$ Å and the triplet state potential for $r_{HH} > 1.669$ Å.

However, the correlation realtionships written above for H_2O hold only for non-linear configurations. The ground state of linear H–O–H is $^1\Sigma_g^+$ and this cannot spatially correlate with $H(^2S) + OH(^2\Pi)$. The spatially allowed correaltion is to $H(^2S) + OH(^2\Sigma^+)$. In other words the H + OH correlation depends strictly on the shape of the triatomic and not just on the three bond lengths treated as independent variables. It is in this sense that the restriction (54) on V_I introduces an approximation to the potential. In effect by making it we are allowing for the spin correlation rules but not spatial correlation rules.

A more thorough analysis shows that to satisfy spatial correlation rules in general one needs to adopt a potential which is not single valued. This is clear if we consider different routes for the dissociation of ground state H_2O into three atoms. If we break one OH bond without passing through a linear structure, to give $OH(^2\Pi)$, and then break the second bond, we obtain three ground state atoms $2\,H(^2S) + O(^3P)$. However, if we first go to a linear structure we obtain the oxygen in an excited state because $OH(^2\Sigma^+)$ dissociates to $H(^2S) + O(^1D)$. Thus starting from one point on the ground state surface we can end up on two different surfaces. The Jahn-Teller conical intersection shown in figure 6 is an example of a double-valued surface.

An n-valued surface can be constructed from the eigenvalues of an $n \times n$ matrix whose elements are continuous functions of the internal coordinates of the molecules. Thus the Jahn-Teller surfaces can be represented by the eigenvalues of a 2×2 matrix such as (23), which are given by expression (24). At the point of intersection the diagonal elements of this matrix are equal and the off-diagonal elements are zero. The LEPS function (30), if extended to allow both signs of the square root, would also represent a two-valued surface. However, a more general attack on the problem in which one attempts to determine the best analytical form of all elements in an $n \times n$ matrix to represent an n-valued surface, appears difficult at the present time.

For this reason single valued functions such as (53) are still of considerable value as a basis for dynamical calculations.

When constructing an analytical potential for a diatomic molecule suitable for spectroscopy or low energy scattering data we are not greatly concerned with its behaviour at large negative values of the displacement from equilibrium. The only requirement is that the potential shall rise steeply in this region. For the triatomic molecule however we are in general interested in large negative displacements from equilibrium. For example, in H_2O the H–H equilibrium separation is 1.51 Å and if the potential deduced from spectroscopic properties of H_2O is to be used to analyse the collision $H_2 + O(^1D)$ then we will require the potential V_I to be valid for values of r_{HH} of the order of – 0.8 Å.

Investigation of an acceptable form of the function V_I, based on an extension of the diatomic potential (45) has shown that an exponential form having the correct rate of decay for positive r rises too steeply for negative r. To overcome this difficulty the following function has been used (*135*).

$$P(r_1, r_2, r_3)\,(1-\tanh\gamma_1 r_1/2)\,(1-\tanh\gamma_2 r_2/2)\,(1-\tanh\gamma_3 r_3/2)\,. \quad (55)$$

We note that $(1-\tanh\gamma r/2)$ behaves as $2\exp(-\gamma r)$ for $r \to \infty$ and approaches the limit 2 as $r \to -\infty$. Physically we are imposing a condition such that if any internuclear distance is small the diatomic potential for that distance dominates the whole.

A 1 : 1 relationship can be established between the coefficients of the polynomial $P(r_1, r_2, r_3)$ in (55) and force constants of the triatomic transferred from the (r_1, r_2, α) to the (r_1, r_2, r_3) set. This leaves open the determination of the parameters $\gamma_1, \gamma_2, \gamma_3$. For H_2O, $\gamma_1 = \gamma_2$ were taken to be equal to the exponent in the OH diatomic potential of the form (45) and γ_3 was treated as a variable parameter which as a minimum requirement satisfied the condition that for the whole potential there were no physically accessible regions lower in energy than the equilibrium configuration.

The final question to be answered is on the form of V_I for the "outer" part of the H_2 potential, that is in the region where the triplet state of H_2 defines the V_{HH}. Unless V_I is changed on passing to this region the overall potential will have a cusp along the line corresponding to the H–H separation of 1.669 Å. To make the potential continuous in all its derivatives along this line is not possible with the functions proposed due to the different analytical forms for the diatomic and interaction parts. One can however make it continuous for a Taylor expansion of the potential to the same order as the highest term in the polynomial in V_I. With this condition and those defined earlier Sorbie and Murrell derived an analytical potential for H_2O and examined contours of this potential for a number of sections in the 4-dimensional space. The only unsatisfactory behaviour was for the linear configurations O–H–H for which the energy appears to be too low.

Fig. 15 shows contours for the insertion of $O(^1D)$ into H_2 in C_{2v} configurations. *Gangi* and *Bader* (*137*) have calculated optimum values of x for several values of y for this reaction and their points which should be on tangents to the contours parallel to the x axis, agree closely in position and energy with the Sorbie-Murrell potential.

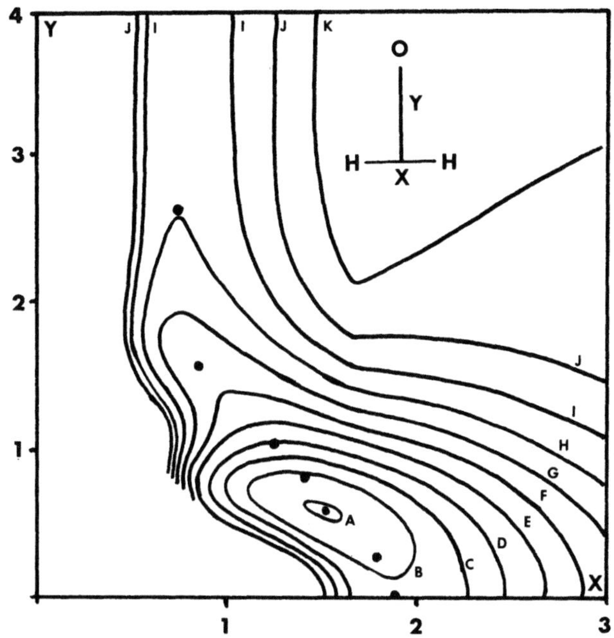

Fig. 15. Contours for the insertion reaction of O(^{1}D) into H_2. Contours are at 1 eV intervals (A = – 10 eV, K = 0).

There is no evidence either from this potential or from calculation to support a substantial barrier for this reaction as has been indicated (*138*).

As a quantitative test of the potential the vibrational eigenvalues have been calculated by a variational method using the program of *Whitehead* and *Handy* (*139*). Table 4 shows the calculated fundamental frequency and one overtone. They are compared with values calculated directly from the Taylor – expanded potential using the same method, and with experimental band heads. The improvement of the Murrell-Sorbie potential over the Taylor expansion is striking. The results on H_2O give some cause for optimism about a procedure in which other polyatomic surfaces are deduced largely from spectroscopic data. The procedure has been applied to other triatomics, notably O_3 and SO_2 with similar success (*140*).

To extend the procedure to larger molecules it would first be necessary to establish that the many-body potential energy can be expressed as a convergent expansion in *n*-body terms. Chemists have adopted this approach by implication for regions of the surface close to equilibrium configurations. The general success of assumptions regarding the additivity of bond energies and the transferability of bond force constants establishes this. However it is not obvious that this approach will work for other regions of the surface, for example at transition states. *Varandas* and *Mur-*

Table 4. Comparison of variationally calculated excitation energies (cm^{-1}) for H_2O with the spectroscopic band heads.

	Sym. stretch	Antisym. stretch	bend	Overtone bend
Murrell-Sorbie (*135*)	3671	3759	1596	3148
Hoy-Mills-Strey (*131*)	3718	3821	1600	3167
Experiment	3652	3756	1595	3151

rell (*141*) have obtained a potential for H_4 based on a many-body expansion which reproduces the main features of the surface that are known. This has been extended to H_5 and H_6 on the assumption that the expansion can be terminated at the 4-body term, but it is clear that for regular hexagons of H_6 one requires a small repulsive 5-body term to give the correct stability of the molecule relative to 3 H_2.

Conclusion

I hope that this review illustrates the enhanced activity in the field of polyatomic potential energy surfaces which has occurred over the past five years. Advances are being made on several fronts and I believe that major gains in the future will arise by combining knowledge from several of these. The Faraday Discussion on Potential Energy Surfaces, which was held in September 1976, had as its objective the gathering together of specialists in different fields with a common interest in potential energy surfaces and it is likely that this meeting will act as a further stimulus in the field.

References

1. *Born, M., Oppenheimer, J.R.:* Ann. Phys., *84*, 457 (1927).
2. *Burrau, O.:* Det. Kgl. Danske Vid. Selskab., *7*, 1 (1927).
3. *Heitler, W., London, F.:* Z. f. Phys., *44*, 455 (1927).
4. *London, F.:* Z. Electrochem., *35*, 552 (1929).
5. *Jepsen, D.W., Hirschfelder, J.O.:* J. Chem. Phys., *32*, 1323 (1960).
6. *Dalgarno, A., McCarroll, R.:* Proc. Roy. Soc., A*237*, 383 (1956); A*239*, 413 (1957).
7. *Fröman, A.:* J. Chem. Phys., *36*, 1490 (1962).
8. *Bunker, P.R.:* J. Mol. Spect. *42*, 478 (1972).
9. *Howard, B.J., Moss, R.E.:* Molec. Phys., *19*, 433 (1970).
10. *Born, M., Huang, K.:* Dynamical Theory of Crystal Lattices, Oxford, 1951.

10[a]. *Englman, R.:* Phys. Rev., *129*, 551 (1963); The Jahn-Teller effect in molecules and crystals, Wiley (1971).

11. *Kolos, W., Wolniewicz, L.:* J. Chem. Phys., *41*, 3674 (1964).
12. *Mandl, F.:* Proc. Phys. Soc., *87*, 871 (1966).
13. *O'Malley, F.:* Phys. Rev., *162*, 98 (1967).
14. *Taylor, H.S.:* Adv. Chem. Phys., *18*, 91 (1970).
15. *Dressler, K., Miescher, E.:* Astrophys. J., *141*, 1266 (1965).
16. *Felenbok, P., Lefebvre-Brion, H.:* Can. J. Phys., *44*, 1677 (1966).
17. *Rocin, Y.V., Dalmany, N., Romand, J.:* J. Mol. Spec., *22*, 154 (1967).
18. *Lagerquist, A., Miescher, E.:* Can. J. Phys., *44*, 1525 (1966).
19. *Marcus, R.A.:* J. Chem. Phys., *45*, 4493 (1966).
20. *Light, J.C.:* Adv. Chem. Phys., *19*, 1 (1971).
21. *Wilson, E.B., Decius, J.C., Cross, P.C.:* Molecular Vibrations, McGraw-Hill 1955.
22. *Fukui, K.:* J. Phys. Chem., *74*, 4161 (1970).

24. *Gimsworth, F.R., Spendley, S., Hext, G.R.:* Technometrics *4*, 441 (1962).

23. *Nelder, J.A., Mead, R.:* Computer J., *7*, 308 (1964).

24. *McIver, J.W., Komornicki, A.:* Chem. Phys. Letters, *10*, 303 (1971).
25. *Davidon, W.G.:* Variable Metric Method for Minimization, A.E.C. Res. and Dev. Rep.
26. ANL 5990 (Rev.).
27. *Fletcher, R., Powell, M.J.P.:* Computer J., *6*, 163 (1963).
28. *Murtagh, B.A., Sargent, R.N.H.:* Computer J., *13*, 185 (1970).
29. *Powell, M.J.P.:* Computer J., *7*, 155 (1964).
30. *Garton, D., Sutcliffe, B.T.:* Specialist Periodical Reports, Theoretical Chemistry, vol. I, Chemical Society, 1974.
31. *Dewar, M.J.S., Kirschner, S.:* J. Amer. Soc., *93*, 4291 4292 (1971).
32. *Neumann, von J., Wigner, E.P.:* Z. Physik, *30*, 467 (1929).
33. *Teller, E.:* J. Phys. Chem., *41*, 109 (1937).
34. *Naqvi, K.R., Byers-Brown, W.:* Int. J. Quant. Chem., *6*, 271 (1972).
35. *Languet-Higgins, H.C.:* Proc. Roy. Soc. hand., A*344*, 147 (1975).
36. *Jeffreys, H., Jeffreys, B.S.:* Methods of Mathematical Physics 2nd Ed. p. 140 (1972).

36[a]. *Landau, L.D., Lifshitz, E.M.:* Quantum Mechanics, Pergamon Press p. 265 (1959).

37. *Herzberg, G., Longuet-Higgins, H.C.:* Disc. Far. Soc., *35*, 77 (1963).
38. *Naqvi, K.R.:* Chem. Phys. Letters, *15*, 634 (1972).
39. *Jahn, H.A., Teller, E.:* Proc. Roy. Soc., *161A*, 220 (1937).
40. *Murrell, J.N.:* Chem. Comm. 1044 (1972).
41. *Longuet-Higgins, H.C., Öpik, U., Price, M.H.L., Sack, R.A.:* Proc. Roy. Soc., A*244*, 1 (1958).
42. *Herzberg, G., Teller, E.:* Z. Physik. Chem., *B21*, 410 (1933).
43. *Renner, R.:* Z. Physik, *92*, 172 (1934).
44. *Öpik, U., Pryce, M.H.L.:* Proc. Roy. Soc., *A 238*, 425 (1957).
45. *Bader, R.F.W.:* Molec. Phys., *3*, 137 (1960).
46. *Bader, R.F.W.:* Can. J. Chem., *40*, 1164 (1962).

47. *Boer, den D. H. W., Longuet-Higgins, H. C.:* Molec. Phys., *5*, 387 (1962).
48. *Nicholson, B. J., Longuet-Higgins, H. C.:* Molec. Phys., *9*, 461 (1965).
49. *Hoffman, R., Woodward, R. B.:* Acc. Chem. Res., *1*, 17 (1968).
50. *Bartell, L. S.:* J. Chem. Educ., *45*, 754 (1963).
51. *Pearson, R. G.;* J. Amer. Chem. Soc., *91*, 1252 (1969); *ibid. 91*, 4947 (1969).
52. *Walsh, A. D.:* J. Chem. Soc., 2260, 2266, 2296, 2301 (1953).
53. *Woodward, R. B., Hoffmann, R.:* J. Amer. Chem. Soc., *87*, 395 (1965).
54. *Longuet-Higgins, H. C., Abrahamson, E. W.:* J. Amer. Chem. Soc., *87*, 2045 (1965).
55. *Pearson, R. G.:* Acc. Chem. Res., *4*, 152 (1971).
56. *Silver, D. M., Stevens, R. M.:* J. Chem. Phys., *59*, 3378 (1973).
57. *Niki, T., Mains, G. J.:* J. Phys. Chem., *76*, 3538 (1972).
58. *Bauer, S. H., Lederman, D. M., Resler, E. L., Fisher, E. R.:* J. Chem. Kin., *5*, 93 (1973).
59. *Durmaz, S., Murrell, J. N., Pedley, J. B.:* J. Chem. Soc. Farady II, *69*, 1370 (1973).
60. *Clynne, M. A. A., Thrush, B. A.:* Trans. Faraday Soc., *57*, 69 (1961).
61.. *Murrell, J. N., Laidler, K.:* Trans. Faraday Soc., *64*, 371 (1968).
62. *Murrell, J. N., Pratt, G. L.:* Trans. Faraday Soc., *66*, 571 (1970).
63. *Stanton, R. E., McIver, J. W.:* J. Amer. Chem. Soc., *97*, 3632 (1975).
64. *Stuckelberg, E. C. G.:* Helv. Phys. Acta., *5*, 369 (1932).
65. *Miller, W.H., George, T. F.:* J. Chem. Phys., *56*, 5637 (1972).
66. *Lin, Y. W., George, T.F., Morokuma, K.:* J. Chem. Phys., *60*, 4311 (1973).
67. *Morokuma, K., George, T.F.:* J. Chem. Phys., *59*, 1959 (1973).
68. *Kolos, W., Wolniewicz, L.:* J. Chem. Phys., *49*, 404 (1968); *51*, 1417 (1969).
69. *Dunning, T.H., Pitzer, R. M., Aung, S.:* J. Chem. Phys., *57*, 5044 (1972).
70. *Ermler, W. C., Rosenberg, B. J., Kern, C. W., Shavitt, I.:* Communicated to the March 1974 meeting of the American Physical Society.
71. *Benedict, W.S., Gailer, N., Plyler, E. J.:* J. Chem. Phys., *24*, 1139 (1956).
72. *Hoy, A. R., Mills, I. M., Strey, G.:* Molec. Phys., *24*, 1265 (1972).
73. *Conway, H., Bruner, B. L.:* J. Chem. Phys., *42*, 4047 (1965); *47*, 921 (1967).
74. *Shavitt, I., Stevens, R. M., Minn, F. L., Karplus, M.:* J. Chem. Phys., *48*, 2700 (1968); *49*, 4048 (1968).
75. *Liu, B.:* J. Chem. Phys., *58*, 1925 (1973).
76. *Conroy, H.:* J. Chem. Phys., *51*, 3797 (1969).
77. *Csizmadia, I. G., Kari, R. E., Polanyi, J. C., Roach, A. C., Robb, M. A.:* J. Chem. Phys., *52*, 6205 (1970).
78. *Bauschlicher, C. W., Jr., O'Neil, S. V., Preston, R. K., Shaeffer III, H. F.:* J. Chem. Phys., *59*, 1286 (1973).
79. *Gordon, M. D., Secrest, D.:* J. Chem. Phys., *52*, 120 (1970); ibid *53*, 4408 (1970).
80. *Shaeffer III, H. F., Wallach, D., Bender, C. F.:* J. Chem. Phys., *56*, 1219 (1972).
81. *Lester, Jr., W.A.:* J. Chem. Phys. *53*, 1511 (1970); ibid *54*, 3171 (1971); ibid *57*, 3028 (1972).
82. *Kutzelnigg, W., Staemmler, V., Hoheisel, C.:* Chem. Phys., *1*, 27 (1973).
83. *Bender, C. F., Pearson, P.K., O'Neil, S. V., Shaeffer III, H. F.:* Science, *176*, 1412 (1972); J. Chem. Phys., *56*, 4626 (1972).
84. *O'Neil, S. V., Pearson, P.K., Shaeffer III, H. F., Bender, C. F.:* J. Chem. Phys., *58*, 1126 (1972).
85. *Balint-Kurti, G. A.:* Potential Surfaces for Chemical Reactions, Advances in Molecular Scattering, Ed. K. P. Lawley.
85[a]. *Bader, R. F. W., Gangi, R. A.:* Specialist Periodical Report, Theoretical Chemistry, vol. 2, Chemical Society, London (1975).
86. *Lowdin, P.O.:* Phys. Rev., *97*, 1474 (1955).
87. *Bender, C. F., Davidson, E. R.:* J. Phys. Chem., *70*, 2675 (1966).
88. *Davidson, E. R.:* Adv. Quant. Chem., *6*, 235 (1972).
89. *Das, G., Wahl, A. C.:* J. Chem. Phys., *56*, 3532 (1972).
90. *Sato, S.:* J. Chem. Phys., *23*, 592, 2465 (1955).
91. *Polanyi, J. C., Rosner, S. D.:* J. Chem. Phys., *38*, 1028 (1955).

92. *Muckerman, J.T.:* J. Chem. Phys., *54*, 1151 (1971).
93. *Wilkins, R.L.:* J. Chem. Phys., *57*, 912 (1972); *58*, 2326 (1973).
94. *Kuntz, P.J., Nemeth, E.M., Polanyi, J.C., Rosner, S.D., Young, C.E.:* J. Chem. Phys., *44*, 1168 (1966).
95. *Moffitt, W.:* Proc. Roy. Soc., A*210*, 245 (1951).
96. *Ellison, F.O.:* J. Amer. Chem. Soc., *85*, 3540 (1963).
97. *Balint-Kurti, G.A., Karplus, M.:* in "Orbital Theories of Molecules and Solids", Clarendon Press, Oxford, 1974.
98. *Porter, R.N., Karplus, M.:* J. Chem. Phys., *40*, 1105 (1964).
99. *Murrell, J.N., Harget, A.J.:* Semi-Empirical SCF MO theory of molecules, Wiley (1972).
100. *Pople, J.A., Santry, D.P., Segal, G.A.:* J. Chem. Phys., *43*, S129 (1965).
101. *Baird, N.C., Dewar, M.J.S.:* J. Chem. Phys., *50*, 1262 (1968).
102. *Bingham, R.C., Dewar, M.J.S., Lo, D.H.:* J. Amer. Chem. Soc., *97*, 1285 (1975).
103. *Dewar, M.J.S., Kirschner, S.:* J. Chem. Soc., Chem. Commun., 463 (1975).
104. *Kolos, W., Wolniewicz, L.:* J. Chem. Phys., *49*, 404 (1968).
105. *Certain, P.R., Hirschfelder, J.O., Kolos, W., Wolniewicz, L.:* J. Chem. Phys., *49*, 24 (1968).
106. *Power, E.A.:* Chapter 4 in "Intermolecular Forces", Adv. Chem. Phys., *12*, 167 (1967), Ed. J. O. Hirschfelder..
107. *Longuet-Higgins, H.C.:* Proc. Roy. Soc., A*235*, 537 (1956).
108. *Murrell, J.N., Shaw, G.:* J. Chem. Phys., *46*, 1768 (1968).
109. *Kreek, H., Meath, W.J.:* J. Chem. Phys., *50*, 2289 (1969).
110. *Kreek, H., Pan, Y.H., Meath, W.J.:* Molec. Phys., *19*, 513 (1970).
111. *Dalgarno, A.;* Chapter 3 in Intermolecular Forces", Adv. in Chem. Phys., *12*, 143 (1967), Ed. J.O. Hirschfelder.
112. *Margenau, H., Kestner, N.R.:* "Theory of Intermolecular Forces", Pergamon Press (1969).
113. *Murrell, J.N., Shaw, G.:* J. Chem. Phys., *46*, 1768 (1968).
114. *Hirschfelder, J.O., Meath, W.J.;* Chapter 1 in "Intermolecular Forces", Adv. Chem. Phys., *12*, 3 (1967), Ed. J.O. Hirschfelder.
115. *Claverie, P.:* Int. J. Quant. Chem., *5*, 273 (1971).
116. *Bertoncini, P., Wahl, A.C.:* Phys. Rev. Letts., *25*, 991 (1970); J. Chem. Phys., *58*, 1259 (1973).
117. *Stevens, W.J., Wahl, A.C., Gardner, M.A., Karo, A.M.:* J. Chem. Phys., *60*, 2195 (1974).
118. *Conway, A., Murrell, J.N.:* Molec. Phys., *23*, 1143 (1974); *27*, 873 (1972).
119. *Gordon, R.G., Kim, Y.S.:* J. Chem. Phys., *56*, 3122 (1972).
120. *Rae, A.I.M.:* Chem. Phys. Letts., *18*, 574 (1973).
121. *Varshni, Y.P.:* Rev. Mod. Phys., *29*, 664 (1957).
122. *Goodisman, J.:* "Diatomic Interaction potential theory", Volme I, Chapter 2, Academic Press (1973).
123. *Hulbert, H.M., Hirschfelder, J.O.:* J. Chem. Phys., *9*, 61 (1941).
124. *Murrell, J.N., Sorbie, K.S.:* J.C.S. Faraday II, *70*, 1552 (1974).
125. *Rydberg, R.:* Z. Physik, *73*, 371 (1931).
126. *Vanderslice, J.T., Mason, E.A., Maisch, W.G.:* J. Mol. Spect., *3*, 17 (1959).
127. *Dunham, J.S.:* Phys. Rev., *41*, 721 (1932).
128. *Simons, G., Parr, R.G., Finlan, J.M.:* J. Chem. Phys., *59*, 3229 (1973).
129. *Bickes Jr., R.W., Bernstein, R.B.:* Chem. Phys. Lett., *26*, 457 (1974).
130. *Pliva, J., Çihla, Z.:* Colln. Czech Chem. Commun., *23*, 777 (1958).
131. *Hoy, A.R., Mills, I.M., Strey, G.:* Molec. Phys., *24*, 1265 (1972). This paper contains other references to the work of Pliva, Kuchitsu, Morino, Overend and their co-workers who have made important contributions to the field.
132. *Eckart, C.:* Phys. Rev., *47*, 552 (1935).
133. *McKellar, A.R.W., Walsh, H.L.:* J. Chem. Phys., *55*, 595 (1971); Can. J. Phys., *50*, 1458 (1972).
134. *Le Roy, R.L., Van Kranendonk, J.:* J. Chem. Phys., *61*, 4750, (1974).

135. *Sorbie, K. S., Murell, J. N.:* Molec. Phys., *29*, 1387 (1975).
136. *Kolos, W., Wolniewicz, L.:* J. Chem. Phys., *49*, 404 (1968).
137. *Gangi, R. A., Bader, R. F. W.:* J. Chem. Phys., *55*, 5369 (1970).
138. *Herzberg, G.:* 'Electronic Spectra and Electronic Structure of Polyatomic Molecuels', Van Nostrand Reinhold, 439 (1966).
139. *Whitehead, R. J., Handy, N. C.:* J. Molec. Spectrosc., *55*, 356 (1975).
140. *Murrell, J. N. and co-workers:* to be published.
141. *Varandas, A. J. C., Murrell, J. N.:* Disc. Faraday Soc., *62*, in Press.

Optical Electronegativity and Nephelauxetic Effect in Oxide Systems

Application to Conducting, Semi-conducting and Insulating Metal Oxides

J. A. Duffy

Department of Chemistry, University of Aberdeen, Scotland

Table of Contents

1. Introduction

The first intense (Laporte-allowed) electron transfer absorption band of a complexed ion is generally due to the transition of an electron essentially from a ligand orbital to an orbital of the central ion; for an octahedral complex of a transition metal, for example, the transition is from the $(\pi + \sigma)\, t_{1u}$ orbital set to the $(d)\, t_{2g}$ or $(d)\, e_g$. The energy of this absorption band has been correlated, by *Jørgensen* (*1*), with the electronegativity difference between the surrounding ligands and the central ion such that

$$\upsilon^{\text{e.t.}}_{\text{corrected}} = 30{,}000\{\chi_{\text{opt}}(\text{ligand}) - \chi_{\text{opt}}(\text{central ion})\} \qquad (1)$$

where χ_{opt} is the "optical" electronegativity (for halide ligands χ_{opt} equals the Pauling electronegativity) and $\upsilon^{\text{e.t.}}_{\text{corrected}}$ (in wavenumbers) is the frequency maximum corrected for the effects of spin pairing and $10\,Dq$.

The concept of electronegativity involves the sharing of electrons between atoms (or ions) bonded to each other. This sharing of electrons can also be envisaged from the point of view of coordination, that is in terms of the degree of electron donation by ligands to a central ion. This gives rise to an expansion of the outer orbitals of the central ion, owing to central field and symmetry restricted covalency effects, and manifests itself as the "nephelauxetic effect" which can be expressed in terms of the ratio $\beta = B_{\text{complex}}/B_{\text{free ion}}$ where B_{complex} and $B_{\text{free ion}}$ are the (Racah) interelectron repulsion parameters of the central ion, respectively complexed and in the free (gaseous) state; (2) β is usually obtained from *d–d* absorption spectra.

Thus within the coordination sphere there are two distinct types of electronic transition:

(i) electron transfer from ligand to central ion, and
(ii) within energy levels associated essentially with the central ion.

The energies of both of these are affected by the degree of electron donation in the ligand-central ion bond, and it follows therefore that some fairly straightforward relationship between them might be thought to exist. Examination of spectral data, however, does not at first sight seem to reveal any such relationship between optical electronegativity and the nephelauxetic effect. For example, even though bromide and cyanide both have the same χ_{opt} value, 2.8, (3) the nephelauxetic effect in bromo complexes and in cyano complexes differs markedly (e.g. β is 0.40 in $[RhBr_6]^{3-}$ but 0.32 in $[Rh(CN)_6]^{3-}$ (*4*)).

In this paper we deal with complexes in which the donor atom is exclusively oxide(–II). Such complexes include aquo complexes, which are well documented, and oxyanion complexes such as phosphato, borato, etc. which are present in vitreous phosphates, borates, etc. and in anhydrous oxyacids. We therefore are dealing with complexes, of metal ions M^{n+}, comprising a system which can be written schematically as

$$M^{n+} \leftarrow O - X.$$

The atom X is P, B, etc. and there will be other oxide(– II)s attached to it (also, for the H_2O ligand X = 2 H). The atom X plays an important role in determining the chemical properties of a metal or non-metal oxide or oxyanion, and thus it is anticipated that the nature of X would influence the oxide(– II) moiety governing (i) the extent of electron donation to M^{n+}, and (ii) the ease with which an electron may be transfered from oxide(– II) to an orbital essentially owned by M^{n+}. In other words, both the nephelauxetic effect produced by the oxide(– II) and the optical electronegativity are expected to be influenced by atom X.

The parameter which expresses the extent of electron donation in the ligand-to-metal ion bond is Jørgensen's h parameter, (*2*) equal to $(1 - \beta)/k$ where k is a parameter specific to the transition metal ion. h has been assigned a value of unity for aquo complexes and is equal to 2.0 for chloro complexes, 2.3 for bromo complexes, etc. Unfortunately the spectral data of transition metal ions give rather unrefined values of h as far as the purpose of observing nephelauxetic trends in different oxyanion media is concerned. Far more sensitive are the p block metal ions with oxidation number two units less than the Group number (outer configuration: s^2). These ions have u.v. absorption bands arising from the s – p inter-shell transitions $^1S_0 \rightarrow {}^3P_1$ and $^1S_0 \rightarrow {}^1P_1$, and the frequencies of these bands undergo large shifts when the h parameter of the ligand set is changed. It has previously been shown (*5*) for these metal ions that a linear relationship exists between h and the frequency, ν, of the singlet-triplet band such that

$$\nu = \nu_f - m\nu_f h \qquad (2)$$

where $-m\nu_f$ is the slope of the line (m is equivalent to k for transition metal ions) and ν_f is the "free ion" frequency obtained by extrapolating to $h = 0$. (Often the frequency of the singlet-singlet band is at too high an energy for successful detection owing to interfering optical absorption by the medium, but nevertheless existing data indicate that a similar linear relationship exists for this transition.)

It might be expected that the extrapolated frequency, ν_f, would be the same as the experimentally determined frequency obtained from gas discharges, since in the case of *d–d* spectra h is zero when $B_{complex} = B_{free\ ion}$. As seen in Table 1 agreement is quite close for the uni and divalent ions. For Sb^{3+} and Bi^{3+} however there is a large discrepancy and the reason for this is not properly understood. From the experimental free ion frequencies for gaseous Sb^{3+} and Bi^{3+} it would be anticipated that when complexed by ligands with small h parameters (around unity) the frequencies of their complexes would be significantly greater than the experimental values that have previously been assigned to them; the spectral data are always obtained from very dilute solutions and thus the question arises as to the exact identity of the complex species that is present. Often antimony(III) and bismuth(III) exhibit their cationic chemistry in the form of the SbO^+ and BiO^+ ions.(*6*). These (solvated) species probably exist in dilute solutions of Sb(III) and Bi(III) not only in water but in the other oxyanion media such as 100% phosphoric acid and sulphuric acid (*7*) and it is possible they are responsible for the spectral data that have been used for the calculation of m and ν_f.

Table 1. Frequency (cm^{-1}) of $^1S_0 \rightarrow {}^3P_1$ absorption and m value (Eq. (2)) for s^2 ions.

Metal ion	Frequencies		m
	gaseous	extrapolated	
In^+	43,349	44,000	0.10
$T\ell^+$	52,393	55,300	0.14
Sn^{2+}	55,196	55,000	0.18
Pb^{2+}	64,391	60,700	0.20
Sb^{3+}	66,700	52,000	0.2
Bi^{3+}	75,926	56,000	0.23

It is interesting to note that if these spectral data are assumed to arise from four-coordinated SbO^+ and BiO^+ species, then the effective increase in m yields ν_f values of around 60,000 cm^{-1} for Sb(III) and 70,000 cm^{-1} for Bi(III)–values that are significantly closer to the experimental free ion frequencies. The existence of a (coordinated) SbO^+ or BiO^+ ion would also explain the greater band width observed for the $^1S_0 \rightarrow {}^3P_1$ absorption which occurs with Sb(III) and Bi(III), since the lower symmetry of the metalyl species would lift the degeneracy (threefold in cubic symmetry) of the excited state.

It has been pointed out by *Reisfeld* (*8, 9*) that the value of h obtained for a vitreous medium depends upon the particular s^2 ion used for the spectral determination. This variation arises owing to the incomplete separation of h and m and is also observed for transition metal ions. The variation, however, is fairly small, but nevertheless was recognised as an initial difficulty in tackling the problem of basicity in oxide-containing media (*10*). To overcome the problem, the Lewis basicity of an oxide-containing medium was expressed as a *ratio* of h parameters: $h'/h_{O^{2-}}$ where h' is the h value as determined for a medium by a particular s^2 ion and $h_{O^{2-}}$ is the h value for the same s^2 probe ion in an extremely basic oxide medium (for practicable purposes taken as an alkaline earth oxide). This ratio, known as the "optical basicity", Λ, has been determined for many oxide containing media using the probe ions $T\ell^+$, Pb^{2+} and Bi^{3+}. In general there is good agreement in the Λ values for all three ions, and the application of the optical basicity concept to the problems of glass chemistry, molten salts and slags is making gradual headway (*11, 12, 13, 14*).

Of the above three s^2 ions, most data have been obtained for Pb^{2+}. The reason is that this ion gives the best $^1S_0 \rightarrow {}^3P_1$ absorption spectrum: the bismuth(III) spectral band tends to be somewhat broad (see above) and the thallium(I) band often has a shoulder, or even consists of two distinct maxima, an effect which has been attributed possibly to the Jahn-Teller effect in the excited state (*15*). From Eq. (2) and data in Table 1, it is apparent that the h parameter as measured by Pb^{2+} is given by:

$$h = (60{,}700 - \nu)/12{,}100 \qquad (3)$$

where ν is the frequency (cm^{-1}) of the $^1S_0 \rightarrow {}^3P_1$ absorption maximum in the medium. It should be noted that this evaluation of h is somewhat different from that proposed by *Reisfeld* who, in expressing the nephelauxetic ratio, takes not the corrected value of ν_f, but the experimental frequency of the (gaseous) free ion (*8*).

The experimental determination of optical basicity from the $^1S_0 \rightarrow {}^3P_1$ frequency of an s^2 probe ion requires the medium under investigation to be transparent in the u.v. region. Many substances, such as slags, are not u.v. transparent, and in devising a method for obtaining Λ for these materials, a simple formula was sought from data of materials of known optical basicity which would predict Λ. This formula required the assignment of a "basicity moderating parameter", γ, to each of the different types of (non-oxygen) atom in the medium (*16*). In the present context of attempting to relate the nephelauxetic effect and optical electronegativity, it is significant that the γ parameter (which governs optical basicity and hence h) is related to the Pauling electronegativity of the element (see Fig. 1).

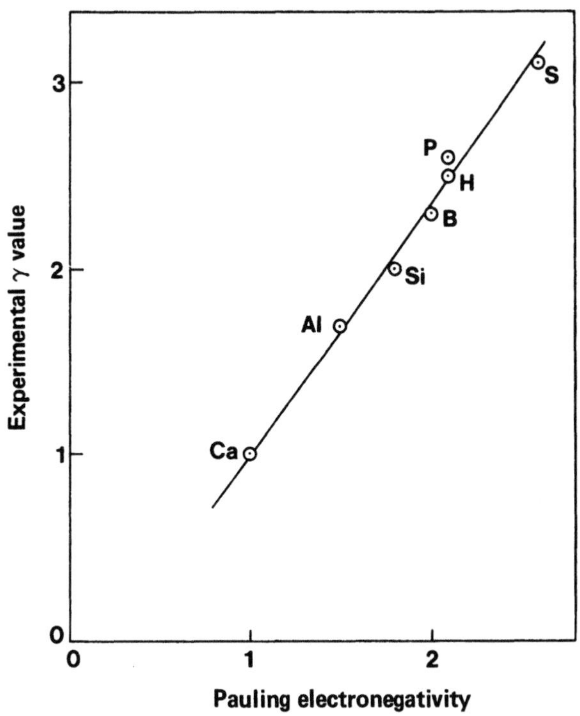

Fig. 1. Relationship between Pauling electronegativity and the experimental basicity moderating parameter, γ, of elements in vitreous oxide systems. Values of γ are calculated from spectroscopic data of the Pb^{2+} probe ion (see *Duffy, J.A., Ingram, M.D.:* J. Chem. Soc. Chem. Comm., 635 (1973)). The straight line fits the equation $\gamma = 1.36\,(x - 0.26)$.

2. *h* Values of Oxyanion Groups

Much of the electron transfer data to be used has previously been obtained and is for metal ions in vitreous media. Glasses usually consist of a network system of oxides linked together by the "network forming" silicon, phosphorus, etc., and when a metal ion such as Cu^{2+}, etc. is introduced into the medium, it is rather difficult to define the coordination sphere containing the metal ion. In principle, the whole of the glass network constitutes the "ligand", but it is probably realistic to limit the extent of complexation by regarding the first coordination sphere as consisting of the mono-oxyanion units coordinated to the metal ion. Thus, for a network arrangement such as

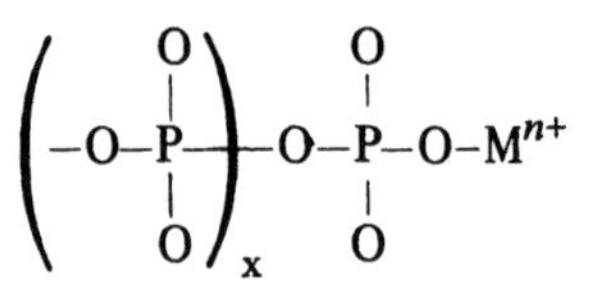

we can limit the coordination sphere to the group

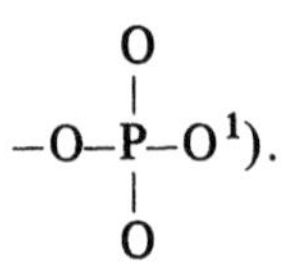

However in doing this, we still have to regard the phosphate group as being at the end of a chain of other phosphate groups and as distinct from the PO_4^{3-} oxyanion.

It is perhaps unfortunate that the metal ions which have *d–d* spectra suitable for yielding *h*, e.g. Cr^{3+}, Ni^{2+}, generally have χ_{opt} values too low for their $\nu^{e.t.}$ frequencies to lie in the accessible part of the u.v. region. (Electron transfer data have been obtained for Fe^{3+} in vitreous media, and although in principle *h* can be obtained from the $^6S - {}^4G$ interval, the weakness of absorption leads to difficulties concerning the probe ion/glass concentration.) However this lack of experimental correspondence would only have been of limited use owing to the restricted sensitivity of transition metal ions towards changes in *h* which was mentioned earlier.

The problem of slight variation of *h* with metal ion has also been mentioned, but this is not serious because in the correlation we are seeking, it is *changes* in *h* which are important. Our present strategy is not to use the experimental values of *h* but to use "ideal" *h* parameters calculated from the constitution of the oxyanion medium and appropriate γ parameters, along the same lines as in the calculation of "ideal" optical basicity (*16*). (It is also possible to evaluate an ideal *h* parameter for the oxide

[1]) In the oxyanion groups, a bridging oxygen is denoted –O– and a terminal –O; the dashes are not meant to imply bond order.

donor atom, but since electron transfers spectra involve an electron from a π molecular orbital of the oxyanion group, it is more appropriate to use h relevant to the group.) The formula in Reference (*16*) can be rearranged to give

$$h = h_{0^{2-}} \left[1 - \frac{z_X}{n_X |z_0|}\left(1 - \frac{1}{\gamma_X}\right)\right] \tag{4}$$

where z is the oxidation number of atom X, $|z_0|$ that of oxide (i.e., 2) and n_X is the number of oxides "owned" by atom X in the oxyanion unit

Table 2. Values of h and χ_{opt} for groups in various oxide media. (These groups are the effective ligands when metal ions are dissolved.)

Medium	Group	h [a]	χ_{opt}
Water	H_2O	1.02 [b]	3.50
Phosphoric acid	$H_2PO_4^-$	1.22	3.34
Phosphate glass (acidic)	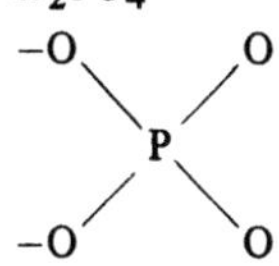	1.28	3.31
Phosphate glass (basic)	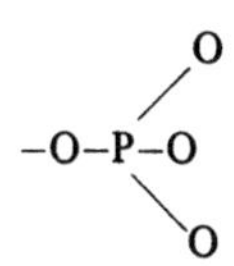	1.46	3.22
Borate glass	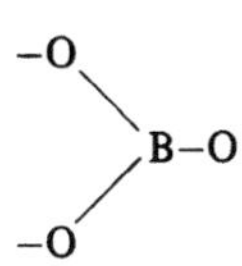	1.46	3.15
$Na_4B_2O_5$ glass	$B_2O_5^{4-}$	1.68	2.90
Silicate glass	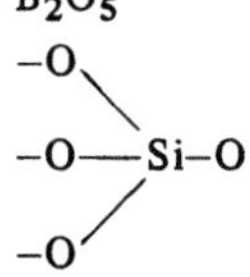	1.48	3.17

[a]) h is calculated from equation (4) except for proton containing groups when the equation

$$h = 2.56 \left\{1 - \frac{z_X}{2 n_X}\left(1 - \frac{1}{\gamma_X}\right) - \frac{0.6}{2 n_H}\right\}$$

is used (see Ref. (*16*)); values of γ are: H, 2.50; P, 2.50; B, 2.36; Si, 2.09 (*16*).

[b]) This value, rather than unity, is assigned for consistency.

(e.g. for 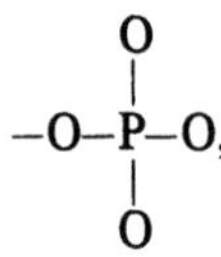 $n = 3\frac{1}{2}$) ;

$h_{O^{2-}}$ is the parameter ppropriate to the ionic O^{2-} ion (taken as the medium CaO) and equals 2.56 (*10*). The values of γ (given in the footnote to Table 2) have been adjusted very slightly from the experimental so that they themselves are idealised to fit exactly the straight line in Fig. 1.

In Table 2 are the h values calculated for the groups which (from the glass constitution) preponder in each glass. It is assumed that it is these groups which coordinate when metal ions are introduced into the glass and a recent study of the Na_2O–P_2O_5 vitreous system strongly supports this view (see below).

3. χ_{opt} Values of Oxyanion Groups

Phosphate media: In the Na_2O–P_2O_5 system, glasses can be produced over a wide range of composition. A critical region is where the $Na_2O:P_2O_5$ ratio is around unity, since increasing the Na_2O content brings about large scale conversion of the preponderant

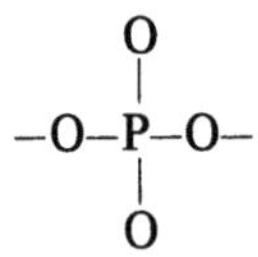

(bridging) group to the

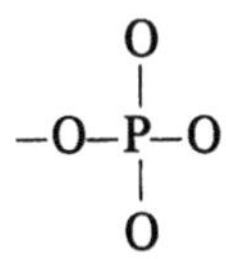

(terminal) group. The calculated increase of 0.18 in h (see Table 2) accompanying this conversion is observed experimentally in the red shift of the $^1S_0 \rightarrow {}^3P_1$ frequencies of Tl^+, Pb^{2+} and Bi^{3+}. At the same time there is a red shift in the electron transfer bands of Cu^{2+}, Fe^{3+} and Mn^{3+} indicating a decrease in optical electronegativity of nearly 0.1 unit (*17*).

Spectral data of transition metal ions in 100% phosphoric acid (*18*) indicate that the nephelauxetic effect is slightly greater than when water is the solvent (e.g. in 100% phosphoric acid and in water, β is respectively 0.87 and 0.89 for Ni^{2+} and 0.75 and 0.79 for Cr^{3+}). Since the calculated value of h for H_3PO_4 is the same as for H_2O (because $\gamma_P = \gamma_H$) these experimental differences are almost certainly owing to some degree of proton loss (i.e. solvolytic dissociation) from the H_3PO_4 ligands, that is the coordination sphere contains a significant proportion of $H_2PO_4^-$ ligands. Indeed, for the Cr^{3+} ion, the β value of 0.75 indicates $h = 1.25$ which is close to the calculated value for $H_2PO_4^-$ in Table 2.

Borate glasses: Increasing the Na_2O content in the Na_2O–B_2O_3 vitreous system brings about a change in the coordination of the boron from three to four and virtually all the oxides retain the role of bridging two boron atoms (*19*). It seems likely that introduction of a metal ion into a sodium borate glass would disrupt the network and cause a four-coordinate

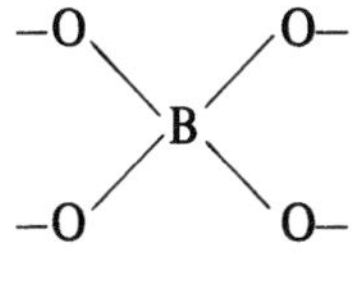

unit to generate a three-coordinate

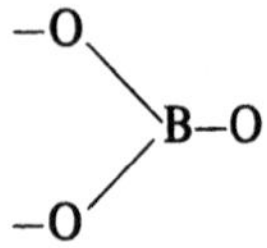

unit and the terminal oxide of this unit would then be able to coordinate with the metal ion. (However even if this did not occur and the coordinating units were instead

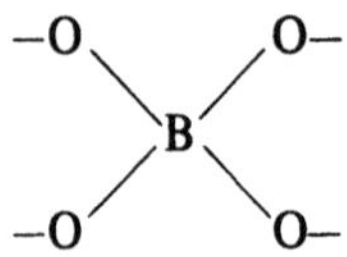

units, it would not matter since the *h* value (Eq. (4) is the same for both.)

The ultra-high alkali oxide borate glass (composition 2 $Na_2O:B_2O_3$) (*20*) does not appear to have a network system but contains discrete $B_2O_5^{4-}$ anions (similar to an acetate glass which contains $CH_3CO_2^-$ ions or a nitrate glass which contains NO_3^- ions (*21*). Accordingly the ligand is taken as $B_2O_5^{4-}$.

Silicate glasses: Owing to the difficulties associated with the high temperatures needed for producing silicate glasses, there are little spectroscopic data available. Electron transfer spectra have been obtained for Cu^{2+}, Fe^{3+} and Mn^{3+} in $Na_2O–SiO_2$ (3:7) glass, and in this glass the preponderant silicate group is

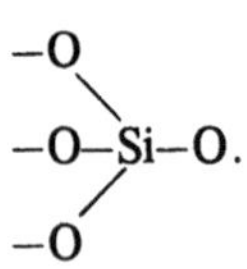

4. Correlation Between Nephelauxetic Effect and Optical Electronegativity

Electron transfer data for transition metal ions in glass and related media have been obtained mainly for Cu^{2+}, Fe^{3+}, Mn^{3+} and, to a lesser extent, Cr^{3+}. In general, the optical electronegativities calculated for the various media agree well regardless of the metal ion. However, the Cu^{2+} ion usually gives the best defined absorption bands in the u.v. region. Thus the values of χ_{opt} in Table 2 are those based upon data for Cu^{2+}. (*17, 18, 20*).

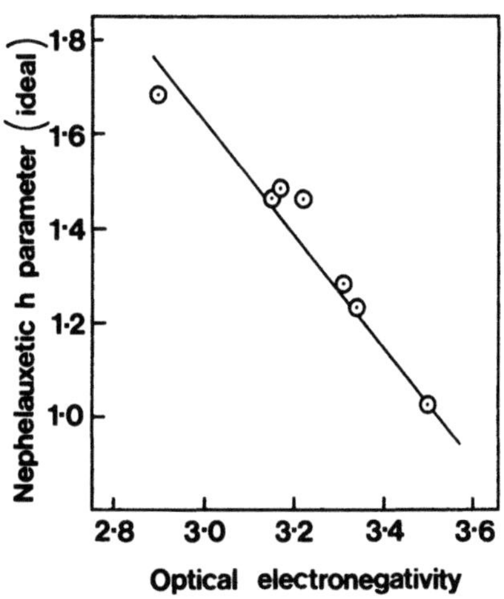

Fig. 2. Plot of ideal *h* parameter against optical electronegativity of oxy-groups in Table 2.

When these values of χ_{opt} are plotted against the *h* value for each oxyanion unit, together with the values of $\chi_{opt} = 3.5$ and $h = 1.02$ for water, the relationship is close to linear – see Fig. 2. The equation for the straight line in Fig. 2 has been drawn to pass through the appropriate point for water and is given by

$$\chi_{opt} = 4.30 - 0.80\,h \qquad (5)$$

The correlation in Fig. 2 does not include data for sulphate media. The SO_4^{2-} ion is predicted to have $\chi_{opt} = 3.3$ (0), while for the HSO_4^- ion (which has $h = 1.08$ from the equation in the footnote to Table 2) $\chi_{opt} = 3.4$ (4). The former value may be compared with the experimentally determined χ_{opt} of 3.1 (0) for $NaHSO_4$–$KHSO_4$ (1:1) glass and K_2SO_4–$ZnSO_4$ (2:3) glass, (*22*) while the latter value may be compared with the experimental χ_{opt} of 3.2 (1) for 100% sulphuric acid (*23*). If the ligands change from SO_4^{2-} to HSO_4^- in going from the sulphate glasses to sulphuric acid, then it is clear that Eq. (5) correlates moderately satisfactorily with the change in χ_{opt} (predicted and observed increases are 0.14 and 0.11 respectively), but the absolute χ_{opt} values are predicted too high by ca. 0.2 unit. It is difficult to account for this disagreement at present.

The relationship (Eq. (5)) has been obtained on the basis of ideal *h* parameters calculated from Eq. (4). In order to check that the relationship is a real one in terms of experimentally determined spectral data, a similar plot to Fig. 2 was made but using *h* values obtained, through Eq. (3), from experimental $^1S_0 \rightarrow {}^3P_1$ frequencies for Pb^{2+}. (Excluded from the plot is the datum referring to 100% phosphoric acid because it has previously been apparent that whereas transition metal ions undergo solvolytic dissociation (see above), the $^1S_0 \rightarrow {}^3P_1$ frequency corresponds to undissociated H_3PO_4 (*5*).) Comparison of the "experimental" plot (Fig. 3) with Fig. 2 confirms that the relationship of Eq. (5) has an experimental as well as a theoretical basis.

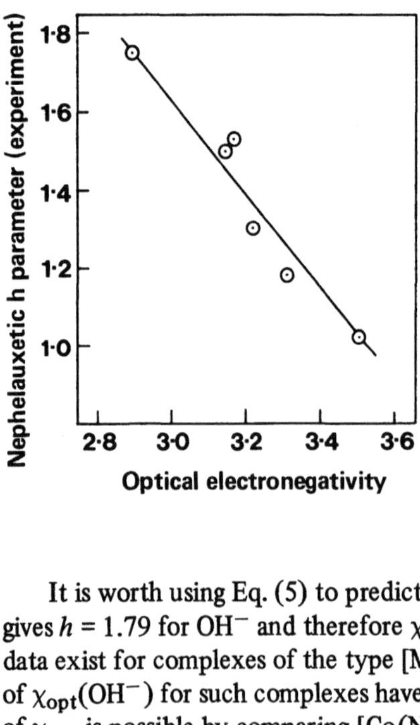

Fig. 3. Similar plot to Fig. 2 but using *h* parameters determined experimentally using Pb^{2+} spectral data and Eq. (*3*).

It is worth using Eq. (5) to predict the optical electronegativity of OH^-. Eq. (4) gives $h = 1.79$ for OH^- and therefore χ_{opt} is predicted as 2·8 (7). No electron transfer data exist for complexes of the type $[M(OH)_6]^{n-}$ and thus no experimental values of $\chi_{opt}(OH^-)$ for such complexes have been determined. Nevertheless, some estimate of χ_{opt} is possible by comparing $[Co(NH_3)_5(H_2O)]^{3+}$ with $[Co(NH_3)_5(OH)]^{2+}$ (*4*); the absorption maximum of the band for the electron transfer is 52,200 cm^{-1} when transfer is from H_2O to Co^{3+} and 34,500 cm^{-1} when transfer is from OH^- to Co^{3+} – a difference of 17,700 cm^{-1} corresponding to a reduction in χ_{opt} of 0.6 unit. With $\chi_{opt}(H_2O) = 3.5$, this reduction yields $\chi_{opt}(OH^-) = 2.9$, in agreement with the prediction of Eq. (5).

5. Ionic, Covalent and Metallic Oxides

The above result for OH^- is encouraging and makes it tempting to use Eq. (5) to obtain a value of χ_{opt} for oxide(– II). Bearing in mind the effect of removing one proton from H_2O, when χ_{opt} falls to 2.9 for OH^-, a value of χ_{opt} which is several 0.1 units below 2.9 is expected for oxide. For CaO h is 2.56 (*10*) and thus χ_{opt} for O^{2-} in CaO is 2.2 (5), though we regard this figure with some caution because the relationship, Eq. (5), might not hold so well at this extrapolation. Nevertheless, even with caution it is apparent that when considering metal oxides such as CaO the χ_{opt} value for oxide(– II) is significantly lower than the value of 3.0 normally quoted for oxide(– II).

χ_{opt} for oxide(– II) is usually obtained from data for oxocomplexes of metals with a high oxidation number, e.g. MnO_4^-, MoO_4^{2-}, OsO_4, and when bound in these complexes it is very different compared with its condition in ionic oxides such as CaO. The oxide ligand can alter its "state" to a much greater degree than any other ligand (*24*) and this depends upon the polarizing nature of the metal (or nonmetal) ion, e.g. Mn^{7+} or Ca^{2+}, with which it is interacting; expressed in other words, the Lewis basicity of oxide can vary over an enormous range, and it is the detection of the change in h accompanying this variation which has allowed Lewis basicity to be expressed numerically on the scale of optical basicity (*10*). Even when oxide is combined with metal ions of high oxidation number, as in oxyanions, it is suspected that its χ_{opt} value is not fixed (*25, 26*). If one goes to the other extreme of chemical combination and considers oxide-metal ion interaction where the oxide is totally ionic, then the O^{2-} ion will spontaneously lose an electron (*27*), and in such a hypothetical compound the value of χ_{opt} for oxide would presumably be zero or even negative. It is interesting to note that if we take χ_{opt} to be zero, Eq. (5) yields $h = 5.4$ and this corresponds to the metal ion having a Pauling electronegativity (calculated from the relationship in Ref. (16)) of 0.6. This is lower than the electronegativity of any of the elements in the periodic table (Cs and Fr have the lowest value of 0.7).

Metal oxides are becoming increasingly important from a technological point of view, especially because of their electrical, magnetic and optical properties. Certain transition metal oxides have extraordinarily high electronic conductivity, e.g. TiO, and indeed are described as "metallic". Most metal oxides however are insulators or semi-conductors, their conductivity being characterized by an energy gap, E_g, which corresponds to the energy difference between the ordinary valence and conduction bands (*28, 29*). This energy difference also manifests itself as the limit of optical transparency, and the absorption edge usually has an energy fairly close (within a few thousand wave-numbers) to the energy gap determined from conduction measurements. Single crystal measurements are very sensitive to the presence of impurities, and reliable data are restricted to a small number of metal oxides, most of which are listed in Table 3.

Table 3. Frequency, $\nu^{e.t.}$, (cm^{-1}) of first Laporte allowed electron transfer band in metal oxides (predicted from optical electronegativity) – comparison with experimental band gap, E_g.

Metal oxide	$\chi_{opt}(M^{n+})$ a)	Effective b) $\chi_{opt}(M^{n+})$	$\nu^{e.t.}$ c)	$\nu^{e.t.}$ d)	E_g e)	Colour
MnO	< 1.9	< 1.4	> 48000	> 27000	23000 – 31000 (2.8 – 3.8)	f
CoO	1.6	1.4	48000	27000	22000 (2.7)	f
NiO	1.8	1.4	48000	27000	30000 (3.7)	f
CuO	2.3 – 2.4	1.9 – 2.0	30000 – 33000	9000 – 12000	5000 (0.6)	Black
ZnO	< 1.4	< 1.4	> 48000	> 27000	ca. 24000 (ca. 3)	White
CdO	1.3 – 1.4	1.3 – 1.4	48000 – 51000	27000 – 30000	19000 (2.3)	Yellow
HgO	1.5	1.5	45000	24000	–	Yellow
V_2O_3	1.9	2.0	30000	9000	800 (0.1)	Metallic
Cr_2O_3	1.8	1.6	42000	21000	27000 (3.4)	f
Mn_2O_3	2.5	2.2	24000	3000	–	Brown
Fe_2O_3	2.5	2.2	24000	3000	16000 (2.0)	Red-brown
Co_2O_3	2.3	1.8	36000	15000	–	Black
Sc_2O_3	ca. 1.2	ca. 1.2	ca. 54000	ca. 33000	44000 – 48000 (5.4 – 6)	White
TiO_2	2.0	2.0	30000	9000	25000 – 27000 (3.05 – 3.2)	White

a) χ_{opt} values for Cu^{2+}, Hg^{2+}, V^{3+}, Cr^{3+}, Fe^{3+}, Co^{3+} from *C.K. Jørgensen:* Halogen Chemistry (Ed. V.Gutmann) Volume 1, London and New York: Academic Press 1967, and Ref. (*25*); for Mn^{3+} from Ref. (*22*); for Sc^{3+}, the estimate of χ_{opt} is from the absorption edge which occurs in 10 M HCl at the u.v. cut-off; (*23*); for the remaining ions χ_{opt} is obtained from spectral data in 100% sulphuric acid (*23*).

b) "Effective χ_{opt}" is derived from the χ_{opt} value by removing corrections for spin pairing effects and 10 *Dq* as described in the Appendix.

c) d) $\nu^{e.t.}$ obtained from Eq. (1) with the operational value of χ_{opt}(oxide(–II)) taken as (*c*) 3.0 and (*d*) 2.3 (see text).

e) Values of E_g are taken from Ref. (*28*) and (*29*) and from C.N.R. Rao and G.V. Subba Rao ("Transition Metal Oxides" – National Bureau of Standards, U.S.A., 1974); also from *Thomas, D.G.;* J. Phys. & Chem. Solids, *15*, 86 (1960) (for ZnO) and H.H. Tippins, ibid., *27*, 1069 (1966) (for Sc_2O_3). Energies are in cm^{-1} and (e.V.).

f) Colour solely due to crystal field absorption.

The application of band theory to account for detailed electrical, optical and magnetic properties has so far had only limited success (*28*). Electronic conduction and optical absorption resulting in the onset of u.v.-visible opaqueness involve the transference of electrons from one ion to another, and it would therefore seem worth applying the principles of optical electronegativity to these problems. Any resulting correlations are expected to be of a much more qualitative nature than results given by applying band theory.

The first problem in considering metal oxides from an optical electronegativity point of view is the assignment of an operational value of χ_{opt} for oxide(– II) in these systems. The results for CaO discussed above indicate that the value of 3.0 is almost certainly too high. Accordingly, in obtaining $\nu^{e.t.}$ (by Eq. (1)) two sets of values were calculated, one with χ_{opt}(oxide(– II)) set at 3.0 and the other based upon the value 2.3 for CaO. These two sets of values are presented in Table 3 so that $\nu^{e.t.}$ can be compared with the energy gap as determined from electrical conductivity and optical absorption edge measurements. It is not worthwhile correcting for the frequency difference between absorption band maximum and its (lower frequency) edge for the purpose of the present comparisons, and it is emphasised that the values of $\nu^{e.t.}$ in Table 3 are *not* serious predictions of absorption maxima for the metal oxides: Eq. (1) strictly only applies to molecules and complex ions and ignores perturbations arising through cooperative effects in binary compounds $M_x L_y$ in which the coordination sphere ML_z is indiscrete. It should be noted that in obtaining the two sets of $\nu^{e.t.}$ the χ_{opt} value for each metal ion is adapted so that the effects of spin pairing and $10\,Dq$ are taken into account; for our purposes, we require the "uncorrected" frequency rather than $\nu^{e.t.}_{corrected}$ (see Appendix).

When the two sets of frequencies are compared with the values of E_g, it would appear that for dipositive transition metal oxides, together with ZnO, CdO and perhaps HgO, the optical electronegativity concept works well provided that the χ_{opt}(oxide(– II)) value chosen is 2.3 rather than 3.0. For tripositive ions, except V_2O_3, it would seem that a value slightly higher than 2.3 would give better agreement between $\nu^{e.t.}$ and E_g, and this is understandable because of the higher oxidation number of + 3 producing a greater "tightening" of the electron charge cloud of oxide, compared with + 2 ions, thereby increasing χ_{opt}(oxide(– II)). The data for Sc_2O_3 and TiO_2 (Table 3) differ from those of the other metal oxides in that the better agreement between $\nu^{e.t.}$ and E_g is obtained by taking χ_{opt}(oxide(– II)) not as 2.3 but as a value closer to 3.0. This is the value obtained on the basis of data for oxyanions of transition metals in their maximum state of oxidation.

When considering the coordination of oxide(– II) to central ions with high oxidation numbers, it is necessary to bear in mind the π antibonding effects upon the central ion *d* orbitals (*30*). The shift in frequency of the first intense (Laporte-allowed) electron transfer band on going from the oxyanion to the halogeno complexes (for which extrapolated frequencies are usually taken) indicates that π antibonding effects must result in decreases in χ_{opt}(central ion) of 0.6 for Mn(VII), 0.5 for Cr(VI) and 0.4 for V(V); neglecting π antibonding effects, the apparent values of χ_{opt}(oxide(– II)) in the oxyanions of these central ions are respectively 3.6, 3.5 and 3.4. At one time a value of 3.2 was taken for χ_{opt}(oxide(– II)), but more recently this has been revised to 3.0.(*26*). With this latter value, χ_{opt}(central ion) is 2.4 for Mn(VII), 2.1 for Cr(VI) and 1.8 for V(V); in using the χ_{opt}(central ion) values of 2.0 for Ti(IV) and ca. 1.2 for Sc(III), no account has been taken of any π antibonding effects, but with these oxidation numbers the effect will be very much smaller than for V(V), Cr(VI) and Mn(VII).(It should also be noted that χ_{opt} for Ti(IV) and Sc(III) apply to octahedral coordination, since we are not dealing with tetrahedral coordination for these two ions.)

6. Classification of Metal Oxides

Attention is now turned to electronegativities other than optical electronegativity (we shall use the symbol x to distinguish from optical electronegativity). When considering oxides that are essentially ionic, non-optical electronegativity scales reveal a large electronegativity difference between the metal and oxygen, e.g. for CaO the Pauling electronegativity difference is $3.5 - 1.0 = 2.5$. Optical electronegativity, having been developed from data relevant to complex ions or molecules within which the bonding is essentially covalent, is at a disadvantage when dealing with ionic oxides. Metal ions in such oxides, by their nature in not forming covalent type complexes, make it very difficult for assigning their optical electronegativity. However, referring again to CaO, the absence of any intensely absorbing u.v. bands until 54, 800 cm^{-1} (*31*) does indicate that the ionic condition corresponds to a large difference also in terms of optical electronegativity.

If we compare CaO with oxides such as CrO_3 or Mn_2O_7 we are going from the essentially ionic to the essentially covalent type of oxide. Although *Pauling* or other (nonoptical) electronegativities are not known for Cr(VI) and Mn(VII), it is generally agreed that increased covalency is accompanied by a decrease in electronegativity difference. In terms of optical electronegativity this trend, if it does even exist, is much less well defined; for example, while for CrO_4^{2-} the difference in χ_{opt} is 0.7, for MoO_4^{2-} it is as high as 1.5 and for WO_4^{2-}, 1.7. (We shall assume that going from oxyanions to neutral molecules produces no change in χ_{opt}(metal ion).) A feature which would seem to distinguish ionic from covalent in metal-oxygen bonding is the change in the value of χ_{opt}(oxide(– II)) which operates. We have already seen how for CaO a value of around 2.3 seems reasonable in contrast to the value of around 3.0 for V(V), Cr(VI) and Mn(VII). Also Sc_2O_3 and TiO fit χ_{opt}(oxide(– II)) having a value nearer to 3.0 than to 2.3. Thus the trend in going from ionic to covalent in metal oxides would appear to be a large jump in χ_{opt}(oxide(– II)) after CaO, followed by a gentler rise to the region 3.0 – 3.2 (if we assume that π antibonding effects do not provide the *entire* correction to frequency – see above). It is interesting to note that this feature is not just confined to optical electronegativity. For example in the series CaO, MgO, BeO, the Pauling electronegativity of oxygen increases with decreasing ionicity, the values respectively being 3.0, 3.1 and 3.4. This emphasises the warning given by several authors that the electronegativity of an element is not an invariant parameter (see Ref. (*32*) for example).

For non-optical electronegativity scales, decreasing electronegativity difference between the two elements of a binary compound generally leads to covalency if x is greater than roughly 2.0, but to metallic bonding if x is less than roughly 1.8. When considering binary oxygen compounds, the trend from ionic to metallic is precluded because of the high electronegativity of oxygen: the trend ionic to metallic for metal oxides cannot be accommodated on these scales. In contrast, however, optical electronegativity does appear to accommodate this trend. In proceeding from CaO to those metal oxides exhibiting dark colours and conducting or semi-conducting properties (and these would appear to be oxides of transition metal ions containing *d* electrons), the decrease in electronegativity difference arises through an increase in χ_{opt}(metal ion) but at the same time, in contrast to what happens in the ionic-covalent trend, with no jump in the value of χ_{opt}(oxide(– II)).

Thus is would appear that we could classify metal oxides into two series:

(i) ionic to covalent, e.g. CaO, Sc_2O_3, ... Mn_2O_7 ... , (ii) ionic to metallic, e.g. CaO, MnO, ... Mn_2O_3 ... and in both series there is an increase in χ_{opt}(metal ion). However, the increase in χ_{opt}(metal ion) is accompanied in (i) by a sharp increase in χ_{opt}(oxide(– II)) (presumably owing to the increasing polar nature of the isoelectronic d^0 ions as oxidation number increases), and in (ii) by only a slight rise (on account of these metal ions having oxidation numbers of only + 2 and + 3). It is emphasised that no claim is being made of a *quantitative* correlation between optical electronegativity and the degree of metallic bonding: other factors such as crystal structure have a far more important effect upon properties such as conductivity and electronegativity parameters certainly seem to be of lesser importance.

The present discussion has been restricted to oxides of the lighter metals of the periodic table. At present, owing to lack of data, it is difficult to judge whether the classification can be applied to the second and third row transition metals and also it is not clear to which classification *p* block metal oxides belong. Possibly a third classification: metallic to covalent should be included, and this would enable relating oxides to each other in terms of a Ketelaar triangle (*33*).

Appendix

The values of "effective x_{opt}" for the transition metal ions in Table 3 are obtained by taking into account the spectroscopic shifts arising from (i) changes in spin pairing energy (when the electron transfers from oxide to metal ion) and (ii) 10 *Dq* (for metal ions with the filled $(t_{2g})^6\,d$ configuration). These energies (in cm^{-1}) are divided by 30,000 cm^{-1} (see Eq. (1) to give the correction Δx_{opt} (Table 4) which is then subtracted from x_{opt}(metal ion).

Table 4. Data for obtaining Δx_{opt} (Frequencies in cm^{-1}).

Metal Ion	B	Spin pairing energy change	10 *Dq*	Δx_{opt}
Mn^{2+}	795 [a]	8 D/3 = 14,800		0.5
Co^{2+}	760 [b]	4 D/3 = 7,100		0.2
Ni^{2+}	800 [a]	2 D/3 = 3,700	8,800 [a]	0.4
Cu^{2+}			12,000	0.4
V^{3+}	420	– 4 D/3 = – 3,700		– 0.1
Cr^{3+}	480 [a]	2 B = 6,700		0.2
Mn^{3+}	570	7 D/3 = 9,300		0.3
Fe^{3+}	540	8 D/3 = 10,100		0.3
Co^{3+}	440	– D = – 3,000	18,000	0.5

Values of B and 10 *Dq* for metal ions in their oxides obtained
a) from Ref. (*2*),
b) from spectral data of A. E. Cherkashin & F. I. Vilesov, Sov. Phys. – Solid State, *11*, 1068 (1969); unreferenced values are obtained as described in the text.

The changes in spin pairing energy depend upon the Racah parameter for the metal ion in its oxide, and in Table 4 the spin pairing parameter is put equal to 7 *B* (*1*). For certain of the trivalent metal oxides, *B* has not been determined, and in Table 4 it is assumed to be 0.66 of the *B* value for the metal aquo ion, since β for Cr_2O_3 is 0.66 β for $[Cr(H_2O)_6]^{3+}$. The likely errors introduced by this assumption will be negligible for our purposes, since an over or underestimate in *B* even as large as 100 cm^{-1} will give rise to approximately 0.05 difference in x_{opt}.

Experimentally determined values of 10 *Dq* are available for Mn^{2+}, Co^{2+}, Ni^{2+}, and Cr^{3+} in their oxides but not for Cu^{2+} and Co^{3+}. However, data for the former set of ions show that the value of 10 *Dq* in their oxides is very close to that in their aquo ions. Accordingly, 10 *Dq* for Cu^{2+} and Co^{3+} in Table 4 is taken as that of the aquo ions.

References

1. *Jørgensen, C. K.:* Oxidation Numbers and Oxidation States, Chapter 7. Berlin–Heidelberg–New York: Springer-Verlag 1969.
2. Reference (1), Chapter 5.
3. *Lever, A. B. P.:* Inorganic Electron Spectroscopy, p. 245. Amsterdam, London, New York: Elsevier 1968.
4. *Jørgensen, C. K.:* Advances in Chemical Physics, *5*, 64 (1963).
5. *Duffy, J. A., Ingram, M. D.:* J. Chem. Phys., *54*, 443 (1971).
6. *Cotton, F. A., Wilkinson, G.:* Advanced Inorganic Chemistry (Interscience, 1972) 3rd ed. page 392.
7. *Dawson, J. L., Wilkinson, J., Gillibrand, M. I.:* J. Inorg. Nucl. Chem., *32*, 501 (1970).
8. *Reisfeld, R., Boehm, L.:* J. Non-Cryst. Sol. *17*, 209 (1975).
9. *Reisfeld, R.:* Structure and Bonding, *30*, 65 (1976).
10. *Duffy, J. A., Ingram, M. D.:* J. Amer. Chem. Soc., *93*, 6448 (1971).
11. *Duffy, J. A., Ingram, M. D.:* J. Non-Cryst. Sol. *21*, 373 (1976)
12. *Easteal, A. J., Udy, D. J.:* Physics Chem. Glasses, *14*, 107 (1973).
13. *Easteal, A. J., Udy, D. J.:* J. Materials Science, *8*, 17 (1973).
14. *Duffy, J. A., Ingram, M. D., Sommerville, I.:* to be submitted.
15. *McClure, D. S.:* Solid State Physics (Academic Press, New York), *9*, 521 (1959).
16. *Duffy, J. A., Ingram, M. D.:* J. Inorg. Nucl. Chem., *37*, 1203 (1975).
17. *Duffy, J. A., Grant, R. J.:* Physics Chem. Glasses, *16*, 124 (1975).
18. *Duffy, J. A.:* Physics Chem. Glasses, *13*, 65 (1972).
19. *Bray, P. J., O'Keefe, J. G.:* Physics Chem. Glasses, *4*, 37 (1963).
20. *Duffy, J. A.:* Physics Chem. Glasses, *16*, 22 (1975).
21. *Duffy, J. A., Ingram, M. D.:* J. Amer. Ceram. Soc., *51*, 544 (1968); J. Chem. Soc. (A), 2398 (1969).
22. *Duffy, J. A., MacDonald, W. J. D.:* Physics Chem. Glasses, *11*, 1 (1970).
23. *Duffy, J. A.:* unpublished results.
24. *Duffy, J. A., Ingram, M. D.:* J. Inorg. Nucl. Chem., *36*, 43 (1974).
25. *Jørgensen, C. K.:* Orbitals in Atoms and Molecules, Chapter 7, London and New York: Academic Press 1962;
26. *Müller, A., Diemann, E., Jørgensen, C. K.:* Structure and Bonding, *14*, 23 (1973).
27. *Jørgensen, C. K.:* Structure and Bonding, *1*, 234 (1966).
28. *Adler, D.:* Solid State Physics (Academic Press, New York), 21, *1* (1968).
29. *Goodenough, J. B.:* Progress in Solid State Chemistry (Pergamon, Oxford and New York), *5*, 145 (1971).
30. *Carrington, A., Jørgensen, C. K.:* Mol. Phys., *4*, 395 (1961).
31. *Whited, R. C., Walker, W. C.:* Phys. Rev., *188*, 1380 (1969).
32. *Gordy, W., Thomas, W. J. O.:* J. Chem. Phys. *24*, 439 (1956); *Cotton, F. A., Wilkinson, G.:* Advanced Inorganic Chemistry, 3rd ed., p. 116 (Interscience, 1972).
33. *Ketelaar, J. A. A.:* Chemical Constitution, 2nd ed. p. 21. New York: Elsevier, 1958.

Author-Index Volume 1—32

GPSR Compliance

The European Union's (EU) General Product Safety Regulation (GPSR) is a set of rules that requires consumer products to be safe and our obligations to ensure this.

If you have any concerns about our products, you can contact us on ProductSafety@springernature.com

In case Publisher is established outside the EU, the EU authorized representative is:

Springer Nature Customer Service Center GmbH
Europaplatz 3
69115 Heidelberg, Germany

Batch number: 10584556

Printed by Printforce, the Netherlands